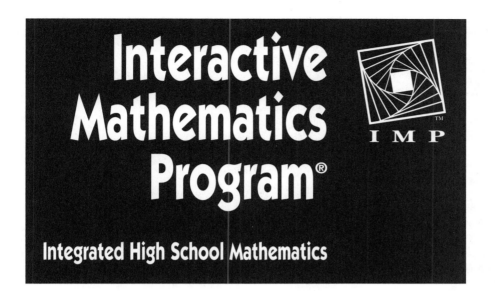

# Interactive Mathematics Program®

**I M P**™

Integrated High School Mathematics

Y E A R **2**

Dan Fendel and Diane Resek

with

Lynne Alper and Sherry Fraser

KEY CURRICULUM PRESS
Innovators in Mathematics Education

This material is based upon work supported by the National Science Foundation under award number ESI-9255262. Any opinions, findings, and conclusions or recommendations expressed in this publication are those of the authors and do not necessarily reflect the views of the National Science Foundation.

Key Curriculum Press
1150 65th Street
Emeryville, California 94608

10 9 8 7 6 5 4       02 01 00 99
ISBN 1-55953-263-7
Printed in the United States of
America.

**Project Editor**
Casey FitzSimons

**Editorial Assistant**
Jeff Gammon

**Production Editor**
Caroline Ayres

**Art Developer**
Jason Luz

**Cover and Interior Design**
Terry Lockman, Lumina Designworks

**Production Manager**
Steve Rogers, Luis Shein

**Production Coordination**
Diana Krevsky, Susan Parini

**Technical Graphics**
Kristen Garneau, Natalie Hill, Greg Reeves

**Illustration**
Tom Fowler, Evangelia Philippidis, Sara Swan,
Diane Varner, Martha Weston, April Goodman Willy

**Publisher**
Steven Rasmussen

**Editorial Director**
John Bergez

**MATHEMATICS REVIEW**
Rick Marks, Ph.D., Sonoma State University,
    Rohnert Park, California

**MULTICULTURAL REVIEWS**
Mary Barnes, M.Sc., University of Melbourne,
    Cremorne, New South Wales, Australia
Edward D. Castillo, Ph.D., Sonoma State University,
    Rohnert Park, California
Joyla Gregory, B.A., College Preparatory School,
    Oakland, California
Genevieve Lau, Ph.D., Skyline College, San Bruno, California
Beatrice Lumpkin, M.S., Malcolm X College (retired),
    Chicago, Illinois
Arthur Ramirez, Ph.D., Sonoma State University,
    Rohnert Park, California

**TEACHER REVIEWS**
Daniel R. Bennett, Kualapuu, Hawaii
Larry Biggers, San Antonio, Texas
Dave Calhoun, Fresno, California
Dwight Fuller, Clovis, California
Daniel S. Johnson, Campbell, California
Brent McClain, Hillsboro, Oregon
Amy C. Roszak, Roseburg, Oregon
Carmen C. Rubino, Lakewood, Colorado
Jean Stilwell, Minneapolis, Minnesota
Wendy Tokumine, Honolulu, Hawaii

# Acknowledgments

Many people have contributed to the development of the IMP curriculum, including the hundreds of teachers and many thousands of students who used preliminary versions of the materials. Of course, there is no way to thank all of them individually, but the IMP directors want to give some special acknowledgments.

We want to give extraordinary thanks to these people who played unique roles in the development of the curriculum.

- **Bill Finzer** was one of the original directors of IMP and helped develop the concept of a problem-based unit.

- **Nitsa Movshovitz-Hadar** suggested the central problem for *Do Bees Build It Best?* and wrote the first draft of that unit.

- **Rich Hemphill** suggested the use of the Alice story as a metaphor for exponential growth and designed several of the activities in *All About Alice.*

- **Matt Bremer** piloted the entire curriculum, did the initial revision of every unit after its pilot testing, and did major work on subsequent revisions.

- **Mary Jo Cittadino** became a high school student once again during the piloting of the curriculum, which gave her a unique perspective on the curriculum.

- **Lori Green** left the classroom as a regular teacher after piloting Year 1 and became a traveling resource for IMP classroom teachers. She has compiled many of her classroom insights in the IMP *Teaching Handbook.*

- **Celia Stevenson** developed the charming and witty graphics that graced the prepublication versions of the IMP units.

In creating this program, we needed help in many areas other than writing curriculum and giving support to teachers.

The National Science Foundation (NSF) has been the primary sponsor of the Interactive Mathematics Program™. We want to thank NSF for its ongoing support, and we especially want to extend our personal thanks to Dr. Margaret Cozzens, Director of NSF's Division of Elementary, Secondary, and Informal Education, for her encouragement and her faith in our efforts.

We also want to acknowledge here the initial support for curriculum development from the California Postsecondary Education Commission and the San Francisco Foundation, and the major support for dissemination from the Noyce Foundation and the David and Lucile Packard Foundation.

Keeping all of our work going required the help of a first-rate office staff. This group of talented and hard-working individuals worked tirelessly on many tasks, such as sending out units, keeping the books balanced, helping us get our message out to the public, and handling communications with schools, teachers, and administrators. We greatly appreciate their dedication.

- Barbara Ford—Secretary
- Tony Gillies—Project Manager
- Marianne Smith—Communications Manager
- Linda Witnov—Outreach Coordinator

## IMP National Advisory Board

We have been further supported in this work by our National Advisory Board—a group of very busy people who found time in their schedules to give us more than a piece of their minds every year. We thank them for their ideas and their forthrightness.

**David Blackwell**
Professor of Mathematics and Statistics
University of California, Berkeley

**Constance Clayton**
Professor of Pediatrics
Chief, Division of Community Health Care
Medical College of Pennsylvania

**Tom Ferrio**
Manager, Professional Calculators
Texas Instruments

**Andrew M. Gleason**
Hollis Professor of Mathematics and Natural Philosophy
Department of Mathematics
Harvard University

**Milton A. Gordon**
President and Professor of Mathematics
California State University, Fullerton

**Shirley Hill**
Curator's Professor of Education and Mathematics
School of Education
University of Missouri

**Steven Leinwand**
Mathematics Consultant
Connecticut Department of Education

**Art McArdle**
Northern California Surveyors Apprentice Committee

**Diane Ravitch** (1994 only)
Senior Research Scholar, Brookings Institution

**Roy Romer** (1992-1994 only)
Governor
State of Colorado

**Karen Sheingold**
Research Director
Educational Testing Service

**Theodore R. Sizer**
Chairman
Coalition of Essential Schools

**Gary D. Watts**
Educational Consultant

We want to thank Dr. Norman Webb of the Wisconsin Center for Education Research for his leadership in our evaluation program, and our Evaluation Advisory Board, whose expertise was so valuable in that aspect of our work.

- David Clarke, University of Melbourne
- Robert Davis, Rutgers University
- George Hein, Lesley College
- Mark St. John, Inverness Research Associates

Finally, we want to thank Steve Rasmussen, President of Key Curriculum Press, Casey FitzSimons, Key's Project Editor for the IMP curriculum, and the many others at Key whose work turned our ideas and words into published form.

Dan Fendel    Diane Resek    Lynne Alper    Sherry Fraser

# *Foreword*

*Is There Really a Difference?* asks the title of one Year 2 unit of the Interactive Mathematics Program (IMP).

"You bet there is!" As Superintendent of Schools, I have found that IMP students in our District have more fun, are well prepared for our State Testing Program in the tenth grade, and are a more representative mix of the different groups in our geographical area than students in other pre-college math classes. Over the last few years, IMP has become an important example of curriculum reform in both our math and science programs.

When we decided in 1992 to pilot the Interactive Mathematics Program, we were excited about its modern approach to restructuring the traditional high school math sequence of courses and topics and its applied use of significant technology. We hoped that IMP would not only revitalize the pre-college math program, but also extend it to virtually all ninth-grade students. At the same time, we had a few concerns about whether IMP students would acquire all of the traditional course skills in algebra, geometry, and trigonometry.

Within the first year, the program proved successful and we were exceptionally pleased with the students' positive reaction and performance, the feedback from parents, and the enthusiasm of teachers. Our first group of IMP students, who graduated in June, 1996, scored as well on PSATs, SATs, and State tests as a comparable group of students in the traditional program did, and subsequent IMP groups are doing the same. In addition, the students have become our most enthusiastic and effective IMP promoters when visiting middle school classes to describe math course options to incoming ninth graders. One student commented, "IMP is the most fun math class I've ever had." Another said, "IMP makes you work hard, but you don't even notice it."

In our first pilot year, we found that the IMP course reached a broader range of students than the traditional Algebra 1 course did. It worked wonderfully not only for honors students, but for other students who would not have begun algebra study until tenth grade or later. The most successful students were those who became intrigued with exciting applications, enjoyed working in a group, and were willing to tackle the hard work of thinking seriously about math on a daily basis.

IMP Year 2 places the graphing calculator and computer in central positions early in the math curriculum. Students thrive on the regular group collaboration and grow in self-confidence and skill as they present their ideas to a large group. Most importantly, not only do students learn the symbolic and graphing applications of elementary algebra, the statistics of *Is There Really a Difference?,* and the geometry of *Do Bees Build It Best?,* but the concepts have meaning to them.

I wish you well as you continue your IMP path for a second year. I am confident that students and teachers using Year 2 will enjoy mathematics more than ever as they experiment, investigate, and discover solutions to the problems and activities presented this year.

*Reginald Mayo*

Reginald Mayo
Superintendent
New Haven Public Schools
New Haven, Connecticut

# Contents

# Appendix: Supplemental Problems . . . . . . . . . . . . . . . . . . . .93

# Is There Really a Difference?

# Days 1–3: Data, Data, Data . . . . . . . . . . . . . . . . . . . . . . .109

# Days 4–9: Coins and Dice . . . . . . . . . . . . . . . . . . . . . . .116

# Days 10-18: A Tool for Measuring Differences . . . . . . . . . . . . 131

# Days 19-23: Comparing Populations . . . . . . . . . . . . . . . . . 158

# Days 24-26: POW Studies . . . . . . . . . . . . . . . . . . . . . . . . 175

# Appendix: Supplemental Problems ...... 180

# Do Bees Build It Best?

## Day 1: Bees and Containers ............... 197

## Days 2-10: Area, Geoboards, and Trigonometry ...... 201

# Days 11-16: A Special Property of Right Triangles . . . . . . . .225

# Days 17-20: The Corral Problem . . . . . . . . . . . .241

# Days 21-27: From Two Dimensions to Three . . . . . . . . .249

## *Cookies*

## *All About Alice*

# *Note to Students*

This textbook represents the second year of a four-year program of mathematics learning and investigation. As in the first year, the program is organized around interesting, complex problems, and the concepts you learn grow out of what you'll need to solve those problems.

## • *If you studied IMP Year 1*

If you studied IMP Year 1, then you know the excitement of problem-based mathematical study, such as devising strategies for a complex dice game, learning the history of the Overland Trail, and experimenting with pendulums.

The Year 2 program extends and expands the challenges that you worked with in Year 1. For instance:

- In Year 1, you began developing a foundation for working with variables. In Year 2, you will build on this foundation in units that demonstrate the power of algebra to solve problems, including some that look back at situations from Year 1 units.

- In Year 1, you used principles of statistics to help predict the period of a 30-foot pendulum. In Year 2, you will learn another statistical method, one that will help you to understand statistical comparisons of populations. One important part of your work will be to prepare, conduct, and analyze your own survey.

You'll also use ideas from geometry to understand why the design of bees' honeycombs is so efficient, and you'll use graphs to help a bakery decide how many plain and iced cookies they should make to maximize their profits. Year 2 closes with a literary adventure—you'll use Lewis Carroll's classic *Alice's Adventures in Wonderland* to explore and extend the meaning of exponents.

## • *If you didn't study IMP Year 1*

If this is your first experience with the Interactive Mathematics Program (IMP), you can rely on your classmates and your teacher to fill in what you've missed. Meanwhile, here are some things you should know about the program, how it was developed, and how it is organized.

The Interactive Mathematics Program is the product of a collaboration of teachers, teacher-educators, and mathematicians who have been working together since 1989 to reform the way high school mathematics is taught. About one hundred thousand students and five hundred teachers used these materials before they were published. Their experiences, reactions, and ideas have been incorporated into this final version.

Our goal is to give you the mathematics you need in order to succeed in this changing world. We want to present mathematics to you in a manner that reflects how mathematics is used and that reflects the different ways people work and learn together. Through this perspective on mathematics, you will be prepared both for continued study of mathematics in college and for the world of work.

This book contains the various assignments that will be your work during Year 2 of the program. As you will see, these problems require ideas from many branches of mathematics, including algebra, geometry, probability, graphing, statistics, and trigonometry. Rather than present each of these areas separately, we have integrated them and presented them in meaningful contexts, so you will see how they relate to each other and to our world.

Each unit in this four-year program has a central problem or theme, and focuses on several major mathematical ideas. Within each unit, the material is organized for teaching purposes into "days," with a homework assignment for each day. (Your class may not follow this schedule exactly, especially if it doesn't meet every day.)

At the end of the main material for each unit, you will find a set of supplementary problems. These problems provide you with additional opportunities to work with ideas from the unit, either to strengthen your understanding of the core material or to explore new ideas related to the unit.

Although the IMP program is not organized into courses called "Algebra," "Geometry," and so on, you will be learning all the essential mathematical concepts that are part of those traditional courses. You will also be learning concepts from branches of mathematics—especially statistics and probability—that are not part of a traditional high school program.

To accomplish your goals, you will have to be an active learner, because the book does not teach directly. Your role as a mathematics student will be to experiment, to investigate, to ask questions, to make and test conjectures, and to reflect, and then to communicate your ideas and conclusions both orally and in writing. You will do some of your work in collaboration with fellow students, just as users of mathematics in the real world often work in teams. At other times, you will be working on your own.

We hope you will enjoy the challenge of this new way of learning mathematics and will see mathematics in a new light.

*Dan Fendel   Diane Resek   Lynne Alper   Sherry Fraser*

# Solve It!

**Days 1-5**

# *Solving Equations and Understanding Situations*

For many people, mathematics means solving equations. You will, in fact, solve lots of equations in this first unit of Year 2. But solving equations makes more sense if those equations describe something meaningful.

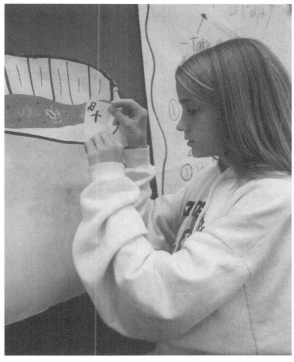

*Tiffany Bock dramatically demonstrates to the class a model for understanding the arithmetic of positive and negative numbers.*

After a look back at Year 1, this unit takes up the challenge of representing situations using algebra and equations. Some of the situations you will study may be familiar from Year 1. The opening days of this unit also include some review of the arithmetic of positive and negative numbers, in the context of a Year 1 Problem of the Week.

# A Year 2 Sampler

You are now beginning Year 2 of the Interactive Mathematics Program. As with Year 1, the curriculum is organized around major units. Here's a brief look at each of these units.

## Solve It!

The first unit of Year 2 focuses on the use of equations to represent problems and on techniques for solving equations, including the basics of algebraic manipulations.

## Is There Really a Difference?

In this unit you'll look at how you might decide whether differences that show up in samples from two populations necessarily represent real differences in the overall populations. You will be introduced to important statistical concepts such as the *null hypothesis* and the *chi-square distribution*.

## Do Bees Build It Best?

In this unit you'll look at the geometry of bees' honeycombs and ask whether this shape is the most efficient. You will work with fundamental ideas about area and volume, learn about the Pythagorean theorem, and continue your work with trigonometry.

*Continued on next page*

# Cookies

The central problem of this unit involves a bakery's decision about how many of each of two kinds of cookies to make. The owners have certain restrictions on oven space, baking time, and so on, and they want to allocate resources in a way that will maximize their profit. You will solve their problem by graphing linear equations and inequalities, solving systems of linear equations with two variables, and reasoning with graphs.

# All About Alice

This unit uses a metaphor from *Alice in Wonderland* to help you explore concepts about exponents and logarithms. It includes material about scientific notation, order of magnitude, and significant digits.

# Homework 1     Math, Me, and the Future

These three writing assignments are designed to help you set the tone for your work in mathematics for this year.

1. Write about how your first year in IMP affected the way you think about mathematics.

2. Write about how the mathematics classes you had before IMP prepared you for Year 1 of IMP.

3. On a separate sheet of paper, write a letter to yourself. Set goals for yourself for this year, imagining that you are your own conscience. You can ask yourself some questions and give yourself some reminders about what you would like to do to succeed in mathematics this year.

Tomorrow in class you will put this letter in an envelope that you address to yourself. Your letter will be delivered to you later in the year.

# Memories of Yesteryear

In this assignment, you will be solving problems based on situations that may be familiar to you. Although you might prefer to solve some of the problems without equations, your assignment is to use variables and equations according to these steps.

- Choose the variable you are going to use in each problem and state what it represents.

- Write an equation, using your variable, that represents the problem.

- Solve the equation and the problem using any method you wish, including guess and check (also known as "trial and error").

1. From *Patterns*

   A chef put several batches of cubes into a cauldron. The first batch contained 27 cubes. The last batch contained 56 cubes. A total of 108 cubes were put into the cauldron.

   How many cubes did the chef throw in that were not part of the first or last batch?

2. From *The Overland Trail*

   Each adult needed 5 yards of shoelace for the trip to California and each child needed 3 yards. A certain family with seven children needed 71 yards of shoelace.

   How many adults were in the family?

*Continued on next page*

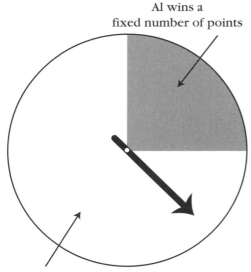

Al wins a
fixed number of points

Betty wins 2 points

### 3. From *The Game of Pig*

Al and Betty were playing a game using the spinner shown here. Betty won 2 points from Al every time the spinner landed on the white section.

Al won a fixed amount from Betty when the spinner landed in the gray section, but you don't know what that amount was. You do know that after 100 games, the results matched the probabilities perfectly, and Al was 25 points ahead of where he started.

How much did Al win from Betty each time the spinner landed in the gray section?

### 4. From *The Pit and the Pendulum*

A group was doing experiments in order to find the period of a pendulum in terms of its length. The group came up with this rule:

If you take the square root of the length of the pendulum (in inches) and multiply that by 0.32, then you will get the number of seconds for one period of the pendulum.

If the period of a certain pendulum is 3.84 seconds, then what is the pendulum's length (based on this rule)?

### 5. From *Shadows*

A person who is 6 feet tall is standing 3 feet from a small mirror that is lying flat on the ground. By looking in the mirror, the person can see the top of a tree that is 15 feet from the mirror.

How tall is the tree?

# Homework 2     Building a Foundation

1. Maisha is planning to build a patio along the back wall of her house, which is 32 feet long. The patio will be rectangular in shape and will fit against the full length of the back wall (so one side of the patio will be 32 feet long).

   The patio will be built out of square tiles that are 1-foot-by-1-foot. Maisha is thinking about this question:

   > If I have 256 tiles to work with, how far out from the wall will the patio extend?

Pretend you are Maisha and do these tasks:

- Make a sketch of the situation.
- Choose the variable you are going to use and state what it represents.
- Write an equation that represents the problem.
- Solve the equation and the problem using any method you wish (including trial and error).

2. Benito is also going to build a patio, but his patio does not have to fit exactly against a wall. In fact, all that Benito has decided is that the patio should be rectangular in shape and should use all of the 144 tiles he has available. (Like Maisha, he is using square tiles that are 1-foot-by-1-foot.)

   Find as many possibilities as you can for the dimensions of Benito's patio. (*Note:* You do not need to go through all the steps you used in Question 1.)

# Homework 3

# You're the Storyteller

In *Memories of Yesteryear,* you started from situations and created equations to fit those situations. In this assignment, you will work in the opposite direction, creating situations that fit the five equations given here. This task has three steps.

• Create a situation.

• Write a question about the situation so that solving the equation will give you the answer to your question. State clearly what the variable in the equation represents in the situation.

• Solve the equation to answer your question.

1. $4a = 12$

2. $r + 5 = 20$

3. $2m + 1 = 11$

4. $\frac{t}{3} = 8$

5. $13 - f = 6$

# Is It a Digit?

There are five empty boxes shown here labeled 0 through 4.

Your task is to put a digit from 0 through 4 *inside* each of the boxes so that certain conditions hold:

- The digit you put in the box labeled "0" must be the same as the number of 0's you use.

- The digit you put in the box labeled "1" must be the same as the number of 1's you use.

- The digit you put in the box labeled "2" must be the same as the number of 2's you use, and so on.

Of course, you are allowed to use the same digit more than once.

You may want to make several copies of the set of boxes in order to try various combinations of digits.

## What Not to Do

Here is an example of an *incorrect* way to fill in the boxes.

| 2 | 3 | 1 | 2 | 2 |
|:-:|:-:|:-:|:-:|:-:|
| 0 | 1 | 2 | 3 | 4 |

This is incorrect for many reasons. For instance, there is a 1 in the box labeled "2," but there is more than one 2 used in the boxes. Similarly, there is a 2 in the box labeled "4," but the number of 4's used is not equal to 2.

# POW 1                                    *A Digital Proof*

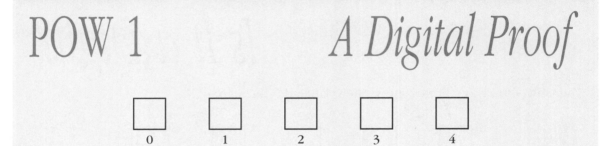

In *Is It a Digit?* you looked for a way to fill in the numbered boxes shown here in a way that fit certain conditions. Your task in this POW is to *prove* that you have all the solutions. (If you haven't yet found a solution, then doing so is also part of your POW.)

## *Write-up*

1. *Problem Statement:* Explain the problem from *Is It a Digit?*

2. *Process:* Based on your notes, describe how you went about finding all of the solutions to *Is It a Digit?* and how you decided that you had them all.

3. *Solutions:* List all solutions you found for *Is It a Digit?* Then write a careful and detailed proof that there are no solutions to *Is It a Digit?* other than those you listed.

4. *Evaluation*

5. *Self-assessment*

(For write-up categories with no specific instructions, use the description in *The Standard POW Write-up.*)

# *The Standard POW Write-up*

Each POW is unique, and so the form of the write-up may vary from one POW to the next. Nevertheless, most of the categories that you will be using for your POW write-ups will be the same throughout the year.

The list below gives a summary of the standard categories for Year 2.

Some POW write-ups will use other categories or require more specific information within a particular category in order to make the write-up more suitable to the POW. But if the write-up instructions for a given POW simply list a category by name, you should use the descriptions below.

*Continued on next page*

# *The Standard POW Write-up Categories*

1. *Problem Statement:* State the problem in your own words. Your problem statement should be clear enough that someone unfamiliar with the problem could understand what it is that you are being asked to do.

2. *Process:* Describe what steps you took in attempting to solve this problem, using your notes to jog your memory. Include steps that didn't work out or that seemed like a waste of time. Complete this part of the write-up even if you didn't solve the problem. And if you got help of any kind on the problem, indicate what form it took and how it helped you.

3. *Solution:* State your solution as clearly as possible. Explain why you think your solution is correct and complete. (If you obtained only a partial solution, give that. If you were able to obtain more general results, include them.)

   Your explanation should be written in a way that will be convincing to someone else—even someone who initially disagrees with your answer.

4. *Evaluation:* Discuss your personal reaction to this problem. For example, you might comment on these questions.

   • Did you consider the problem educationally worthwhile? What did you learn from it?

   • How would you change the problem to improve it?

   • Did you enjoy working on the problem?

   • Was the problem too hard or too easy?

5. *Self-assessment:* Assign yourself a grade for your work on this POW, and explain why you think you deserve that grade.

# Homework 4

# Running on the Overland Trail

For each of the problems here, complete these tasks.

- Choose a variable.

- State clearly what the variable represents.

- Write an equation using your variable that represents the problem.

- Solve both the equation and the problem.

1. If Phillipe had $7 more, he could buy a $30 pair of tennis shoes. How much money does he have?

2. Yolanda jogged 2 miles to a lake, ran twice around the lake, and then jogged 2 more miles home. Altogether she traveled 10 miles. How far is it around the lake?

3. An Overland Trail family is carrying 5 gallons of water per person in its wagon. Unexpectedly, two stragglers join the group. The family figures out that this means there are now only 4 gallons per person. How many people were in this Overland Trail family?
(*Hint:* Take a guess and write down what you should do to see if it's right. Keep doing this until you see a pattern in the arithmetic steps. Then use these steps to come up with an equation.)

# *Lamppost Shadows*

Chantelle and Nelson are members of a volunteer clean-up committee. At the end of the day they are waiting with other volunteers for the shuttle back to the community center.

1. Chantelle is 5 feet tall. She is standing 30 feet from a lamppost that is 25 feet tall. Using *S* to stand for the length of Chantelle's shadow, you can represent her situation using the diagram shown here.

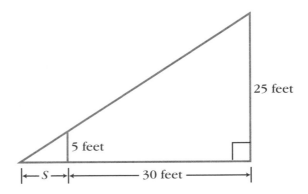

a. Take a guess as to how long the shadow is, *just from looking at the diagram*.

b. Explain, using the diagram, why *S* must fit the equation

$$\frac{S}{S + 30} = \frac{5}{25}$$

c. Try to find a number for *S* that solves this equation. If you can't solve the equation exactly, look for a number that comes close.

d. Compare your answer in Question 1c to your guess in Question 1a. Does your answer in Question 1c seem reasonable?

2. Nelson, who is 6 feet tall, is standing 20 feet from the same lamppost.

a. Draw and label a diagram showing Nelson and his shadow.

b. Write an equation whose solution would give the length of Nelson's shadow.

c. Try to find a number that solves your equation from Question 2b. If you can't solve the equation exactly, look for a number that comes close.

# Homework 5

# 1-2-3-4 Puzzle with Negatives

This assignment is based on *POW 2: 1-2-3-4 Puzzle* from the Year 1 unit *Patterns*. The idea of that problem was to use the digits 1, 2, 3, and 4 once each, along with arithmetic operations, to create expressions with different numerical values. Such expressions are called **1-2-3-4 expressions.** For instance, $1 + (2 + 3) \cdot 4$ is a 1-2-3-4 expression for the number 21.

Unlike the original problem, this assignment involves negative as well as positive integers, so read the instructions carefully.

## The Task

Create as many 1-2-3-4 expressions as you can for each of the numbers from –20 to 20, using the rules outlined here.

*Continued on next page*

# *The Rules*

There is one essential rule for forming 1-2-3-4 expressions.

- You must use each of the digits 1, 2, 3, and 4 exactly once.

The digits can be combined using any of these methods.

- You may use any of the four basic arithmetic operations—addition, subtraction, multiplication, and division (according to the order-of-operations rules).

- You may use exponents.

- You may use radicals or factorials.

- You may juxtapose two or more digits to form a number such as 12.

- You may use parentheses and brackets to change the meaning of the expression.

- You may use a negative sign in front of any of the digits 1, 2, 3, or 4. For example, $-3 \cdot (4^2 - 1)$ is a 1-2-3-4 expression for the number –45. (This method was not included in the original problem.)

*Note:* You may *not* just put a negative sign in front of an entire expression. For example, $-(3 + 4! + 1 - 2)$ is *not* a legitimate 1-2-3-4 expression for –26, even though $-(3 + 4! + 1 - 2)$ is equal to –26 and uses each digit exactly once. You can only put the negative sign in front of an individual digit.

**Days 6-12**

# Keeping Things Balanced

What is an equation? What does it mean to solve an equation? *Homework 6: The Mystery Bags Game* introduces a simple game using a pan balance that will be used throughout the unit as a way to think about these questions.

Over the next few days, you'll review ideas about substitution and order of operations. You'll work with families of algebraic expressions and use functions to represent situations.

For now, trial and error will be one of the main tools for solving equations, but you'll also begin using graphs and seeing how valuable a graphing calculator can be for solving equations.

*Jeff Trubitte and Rafael Pozos created two different outcomes for "POW 2: Tying the Knots."*

# Homework 6

# The Mystery Bags Game

Do you remember the king in the "Bags of Gold" POWs? Well, he doesn't let the gold out of his sight anymore. But it can get very boring watching gold all day, so he has the court jester make up games for him to pass the time.

The game the king loves best is the mystery bags game. First, the jester takes one or more empty bags and fills each bag with the same amount of gold. These bags of equal weight are called the "mystery bags." Next, the jester digs into his collection of lead weights. He takes out his pan balance and places some combination of mystery bags and lead weights on the two pans so that the two sides balance.

The game is to figure out the weight of each mystery bag.

*Continued on next page*

# *Your Task*

The game may sound rather easy, but it can get very difficult for the king. See if *you* can win the mystery bags game in the various situations described here by figuring out how much gold there is in each mystery bag.

Explain how you know you are correct. You may want to draw diagrams to show what's going on. (The picture at the beginning of this assignment shows what the situation in Question 1 might look like.)

1. There are 3 mystery bags on one side of the balance and 51 ounces of lead weights on the other side.

2. There are 1 mystery bag and 42 ounces of weights on one side, and 100 ounces of weights on the other side.

3. There are 8 mystery bags and 10 ounces of weights on one side, and 90 ounces of weights on the other side.

4. There are 3 mystery bags and 29 ounces of weights on one side, and 4 mystery bags on the other side.

5. There are 11 mystery bags and 65 ounces of weights on one side, and 4 mystery bags and 100 ounces of weights on the other side.

6. There are 6 mystery bags and 13 ounces of weights on one side, and 6 mystery bags and 14 ounces of weights on the other side. (The jester could get in a lot of trouble for this one!)

7. There are 15 mystery bags and 7 ounces of weights on both sides. (At first, the king thought this one was easy, but then he found it to be incredibly hard.)

8. The king wants to be able to win easily all of the time, without calling you in. Therefore, your final task in this assignment is to describe in words a procedure by which the king can find out how much is in a mystery bag in any situation.

# Homework 7 You're the Jester

1. Here are some simple equations
   that might have come from
   mystery bags games. Solve
   each equation for *M*, which
   represents the weight of each
   mystery bag.

   a. $M + 16 = 43$

   b. $12M = 60$

   c. $27 + 9M = 90$

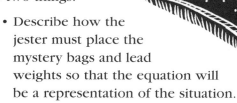

2. The equations in the
   next group are a bit
   more complicated.
   For each equation,
   do two things.

   - Describe how the
     jester must place the
     mystery bags and lead
     weights so that the equation will
     be a representation of the situation.

   - Find the weight of one mystery bag and
     explain how you got the answer.

   a. $5M + 24 = 51 + 2M$

   b. $43M + 37 = 56M + 24$

   c. $12M + 15 = 5M + 62$

3. Make up two equations of your own like those in Question 2. Describe the
   jester's setup for each of your equations, and find the weight of one mystery
   bag in each case.

# Substitution and Evaluation

You often need to find out what the numerical value of a particular algebraic expression would be if you replaced the variable with a number. This happens a lot in the guess-and-check approach to solving equations.

It's useful to identify and name two separate parts of this process of getting numerical values from algebraic expressions.

- **Substitution** is the step of replacing the variable with a number.

- **Evaluation** is the step of getting a single number from the result of the substitution step.

For example, consider the expression $x^2 + 5x - 3$. Suppose you wanted to see what would happen if $x$ were equal to 7.

*Continued on next page*

**Substitution:** In the substitution step, you simply replace each occurrence of the variable with the number 7, as shown here. Recall that 5(7) means 5 *times* 7.

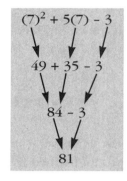

Notice that the number 7 has been placed within parentheses in each case for clarity. This isn't always necessary, but it helps prevent mistakes, such as getting 57 instead of $5 \cdot 7$.

**Evaluation:** The evaluation step turns the numerical expression $(7)^2 + 5(7) - 3$ into a single numerical value. As shown here, you might first replace $(7)^2$ with 49 and 5(7) with 35, then add $49 + 35$ to get 84, and finally subtract 3 from 84 to get the final result.

Here are some useful ways to express this overall process in words.

> "The value of the expression $x^2 + 5x - 3$ for $x = 7$ is 81."

> "Substituting 7 for $x$ in the expression $x^2 + 5x - 3$ gives the value 81."

> "Evaluating $x^2 + 5x - 3$ at $x = 7$ gives 81."

Keep in mind that in doing the evaluation step, you need to follow the rules for order of operations. By convention, we simplify expressions according to this sequence:

1. Parentheses
2. Exponents
3. Multiplication and division (equal priority) from left to right
4. Addition and subtraction (equal priority) from left to right

# *Warning: The Missing Multiplication Sign*

According to the rules for order of operations, we apply an exponent before we multiply. For example, the numerical expression $3 \cdot 7^2$ means $3 \cdot 49$, and not $21^2$.

This rule also governs algebraic expressions, but many errors in the substitution/evaluation process occur because we leave out multiplication signs in algebraic expressions. For example, in the expression $3x^2$, there is a "missing multiplication sign" between 3 and $x^2$. In other words, $3x^2$ is shorthand for $3 \cdot x^2$. Therefore, $3x^2$ means $3 \cdot (x^2)$ and not $(3 \cdot x)^2$. You may find it helpful to insert parentheses or explicit multiplication signs into algebraic expressions in order to be clear about what they mean.

# Homework 8

# Letters, Numbers, and a Story

## Part I: Substitution and Evaluation

Evaluate each of the eight expressions shown here according to these two steps.

- Replace the variable with the value shown, writing the resulting expression in complete detail.

- Compute the numerical value of the expression you get in the first step.

Be sure to insert parentheses or multiplication signs where needed.

*Note:* The instructions in Questions 1 through 8 illustrate some of the many ways by which the process of substitution is described. In each case, you should use both steps.

1. Evaluate $5 + 6q$ at $q = 9$.

2. Find the value of $3z + 20$ when $z = -8$.

3. Get the numerical value of $15 - 4x$ for $x = -1$.

4. Evaluate $3t^2 + 7$ if $t = -2$.

5. What is $-r^2$ when $r = 8$?

6. Find $-z^2$ with $z = -6$.

7. Substitute $k = 3$ into $3 \cdot 2^k + 5$.

8. Evaluate $3x^3 + (4x)^2$ using $x = 5$.

## Part II: Make Up a Story

Create a situation and then write a question about your situation so that solving the equation $4(x + 3) = 40$ will answer your question. Be sure to identify what the variable $x$ represents in your situation. Once you solve the equation, explain what the solution means in terms of your situation.

# Catching Up

The Sawyer family is 120 miles behind the rest of the wagon train and needs to catch up. Both the wagon train and the Sawyers' wagon travel at about 3 miles per hour.

The Sawyer family realizes that in order to make up the difference, they will have to travel more hours each day. They know that the wagon train travels 8 hours each day. Therefore, the Sawyer family decides that they will travel 10 hours every day.

1. a.  How far will the Sawyer family travel in 6 days?

   b.  How far will the wagon train travel during those 6 days?

   c.  Will the Sawyer family catch up in 6 days? Explain your answer.

2. a.  How far will the Sawyer family travel in 15 days?

   b.  How far will the wagon train travel during those 15 days?

   c.  Will the Sawyer family catch up in 15 days?

3. Use *N* to represent a number of days.

   a.  Write an expression for how far the Sawyer family will travel in *N* days.

   b.  Write an expression for how far the wagon train will travel during those *N* days.

   c.  Write an equation that states that the Sawyer family has caught up to the wagon train.

   d.  Solve your equation from Question 3c and interpret the result.

# Homework 9   More Letters, Numbers, and Mystery Bags

## Part I: Substitution and Evaluation

1. As in *Homework 8: Letters, Numbers, and a Story,* evaluate each of these expressions showing these two steps.

   • Replace the variable by the value shown.

   • Compute the numerical value of the resulting expression.

   a. Find the value of $5a^2 + 3a + 4$ for $a = -2$.

   b. Evaluate $-r^3 + 2r^2 + 4r$ when $r = -3$.

   c. What is $(m^2 + 2)(m - 1) - (m - 1)(m^2 + 3)$ if $m = -7$?

   d. Substitute $c = -5$ into the expression $6(c + 4) - 3c(c - 1)$.

   e. Get the numerical value of $(v + 5)(v^2 - 4) - (v - 5)(v^2 + 4)$ at $v = 7$.

   f. Evaluate $y^3 + (2y)^2$ at $y = -5$.

   g. Find $2r^2 - 5r + 9$ with $r = -6$.

*Continued on next page*

2. Make up two substitution examples of your own using these steps.

- Decide what letter to use as the variable.

- Make up an expression using that variable.

- Pick a number to substitute as a value for the variable.

- Substitute the number for the variable and then evaluate the resulting expression.

In one of your examples, substitute a positive number. In the other example, substitute a negative number.

## Part II: More Mystery Bags

3. Solve each of these equations and explain your work using the mystery-bags model.

a. $15M + 43 = 37M + 12$

b. $52x + 19 = 23 + 16x$

c. $5t + 12t + 13 = 8t + 19$

d. $9a + 6 + 3a + 7 = 10a + 21 + 6a$

e. $3r + 4 + 2r = 7 + r + 4r$

# *Back to the Lake*

The following problem is from *Homework 4: Running on the Overland Trail.*

> Yolanda jogged 2 miles to a lake, jogged twice around the lake, and then jogged 2 more miles home. Altogether she traveled 10 miles. How far is it around the lake?

In that particular situation, the distance around the lake was 3 miles.

Well, Yolanda always does a 10-mile jog, and she likes to go 2 miles to the lake and 2 miles back, but she gets tired of always going around the lake twice. Fortunately, Yolanda lives in Minnesota, and several lakes of varying sizes are 2 miles from her home. She would like to be able to choose a lake of the right size depending on how many times she wants to go around.

1. Suppose Yolanda wants to jog 2 miles to a lake, go *four* times around it, jog 2 miles home, and have that be a total of 10 miles. How big a lake should she choose? That is, how far should it be around the lake?

*Continued on next page*

2. Now set up an In-Out table to describe situations like Yolanda's original jog and Question 1. In each case, Yolanda jogs 2 miles to a lake, goes some number of times around it, and jogs 2 miles home, for a total of 10 miles. The *In* for your table should be the number of times Yolanda goes around the lake, and the *Out* should be the distance around the lake.

   a. Use Yolanda's original situation for one row of the table and Question 1 for another. (For instance, for the original situation, the *In* would be 2, because she went around twice, and the *Out* would be 3, because the distance around the lake was 3 miles.)

   b. Create two more rows by choosing two other values for the *In*.

3. Now find a rule for your In-Out table from Question 2. Use *N* for the *In* and *d* for the *Out*, and write *d* as a function of *N*. In other words, write an equation in the form

$$d = \text{an expression involving } N$$

so that Yolanda could substitute any value she wanted for *N* and find the size of the lake she needs.

4. Make a graph based on the In-Out table from Question 2, using your equation from Question 3 to find more points for the graph.

# Homework 10        What Will It Answer?

An important part of understanding what an equation or function means is knowing what types of questions it can be used to answer. For instance, the equation $A = s^2$ gives the area of a square ($A$) as a function of the length of its side ($s$).

The simplest use of this equation is to answer the question "What is the area of the square?" when you know the length of its side. For example, if you know that the length of a side is 5 inches, then you can substitute 5 for $s$ to find that the area of the square is $5^2$ (or 25) square inches.

The equation $A = s^2$ is even more powerful when you realize that it can also be used to answer a different type of question. That is, it can answer the question "What is the length of the side of the square?" if you know the square's area. For example, if you know that the area is 49 square inches, then the equation tells you that $49 = s^2$, which means that the side length must be 7 inches. (Why doesn't a solution of $s = -7$ make sense, even though it fits the equation $49 = s^2$?)

*Continued on next page*

1. The *perimeter* of a square is also a function of the length of a side.

   a. Choose variables and use them to write an equation describing this function. Be sure to state what your variables stand for.

   b. Give specific examples of two different types of questions that your equation can be used to answer. Also give the answers to your questions.

2. A movie theater charges $7 per ticket, and the theater's expenses are $500.

   a. Define appropriate variables and write an equation that gives the theater's profit as a function of the number of tickets sold. (Ignore such factors as the sale of refreshments.)

   b. Give specific examples of two different types of questions that your equation can be used to answer. Also give the answers to your questions.

Many principles in physics can be described in terms of functions. Questions 3 and 4 give two examples of this.

3. If an object is dropped and falls toward the ground, the distance it travels in $t$ seconds is given approximately by the equation $d = 16t^2$, where $d$ is the distance traveled, measured in feet. Come up with two different types of questions that this equation can be used to answer, and give the answers to your questions.

4. Newton's Second Law of Motion states that the force acting on an object ($F$) is equal to the object's mass ($m$) times the object's acceleration ($a$). In other words, the equation $F = ma$ gives the force as a function of mass and acceleration. What types of questions can this function be used to answer?

# POW 2                              *Tying the Knots*

Keekerik is an imaginary land where the people have an interesting three-stage ritual for couples who want to get married. Wandalina and Gerik are in that situation, so they go to the home of Queen Katalana to perform this ritual. Permission for them to marry as soon as they wish depends on the outcome of the ritual.

## Stage 1: Loose Ends Top and Bottom

The queen greets them and reaches into a colorful box to pull out six identical strings for the ritual. The queen hands the strings to Wandalina, who holds them firmly in her fist. One end of each string is sticking out above Wandalina's fist, and the other end of each string is sticking out below her fist.

## Stage 2: The Tops Are Tied

The queen steps to the side, and Gerik is called forward. He ties two of the ends together above Wandalina's fist. Then he ties two other ends above her fist together. Finally, he ties the last two ends above her fist together. The six ends below Wandalina's fist are still hanging untied.

## Stage 3: The Bottoms Are Tied

Now Queen Katalana comes forward again. Although she was watching Gerik, she has no idea which string end below Wandalina's fist belongs to which end above. The queen does the final step. She randomly picks two of the ends below and ties them together, then two more, and finally the last two. So Wandalina now has six strings in her fist, with three knots above and three knots below.

*Continued on next page*

# Will They Be Able to Marry?

Whether Wandalina and Gerik will be allowed to marry right away depends on what happens when Wandalina opens her fist. If the six strings form one large loop, then they will. Otherwise, they will be required to wait and repeat the ritual in six months.

With this in mind, think about these questions.

1. When Wandalina opens her fist and looks at the strings, what combinations of different size loops might there be?

2. What is the probability that the strings will form one big loop? In other words, what are the chances that Wandalina and Gerik will be able to marry right away?

3. What is the probability for each of the other possible combinations?

Although you may want to do some experiments to get some ideas about these questions, your answers for Questions 2 and 3 should involve discussion of the theoretical probability for each result, and not just experimental evidence.

# Write-up

1. *Problem Statement*

2. *Process:* Explain how you worked on this problem, including what experiments you performed and how you kept track of your results.

3. *Solution:* Give the probability for each possible outcome and explain how you determined each probability.

4. *Evaluation*

5. *Self-assessment*

# Homework 11                              Line It Up

Probably the most important single type of function is
the **linear function,** which can be defined as a
function whose graph is a straight line.

1. Consider the function $f$ defined by the
   equation

   $$f(x) = 3x + 2$$

   a. Find each of the values $f(1), f(2),$ and $f(3)$.

   b. Use the results of Question 1a to
      complete this partial In-Out table for the
      function $f$.

   | $x$ | $f(x)$ |
   |-----|--------|
   | 1 | ? |
   | 2 | ? |
   | 3 | ? |

   c. Graph the points from your In-Out table.

   d. Do you think the graph of $f$ is a straight line? In other words, is $f$ a linear
      function? Explain your answer.

2. a. Plot the two points $(2, 3)$ and $(4, 7)$.

   b. Draw a straight line through your two points and find a third point on that
      line.

   c. Make an In-Out table like the one shown
      here, using your point from Question 2b as
      the third row

   | In | Out |
   |----|-----|
   | 2 | 3 |
   | 4 | 7 |
   | ? | ? |

   d. Find a rule for your table in Question 2c.

3. What kind of algebraic expression do you think
   can be used as the rule for a linear function?
   (You might either give some examples or try to provide a general
   description.)

# The Graph Solves the Problem

In Question 2 of *Homework 10: What Will It Answer?* you probably came up with an equation like $p = 7t - 500$ to describe the theater's profit ($p$) in terms of the number of tickets they sold ($t$).

In Question 3 of that assignment, you were given the equation $d = 16t^2$ to describe $d$, the distance an object has fallen (in feet), in terms of $t$, the time elapsed (in seconds).

Use these two equations and the graphing feature of your graphing calculator to answer these questions.

1. a. The theater made a profit of $277 on yesterday's show. How many tickets were sold?

   b. Three hundred people bought tickets for today's show. How much profit did the theater make?

2. a. An object is dropped off the roof of a very tall building. How long will it take for the object to fall 200 feet?

   b. How far will the object have gone if it falls for 6 seconds?

*Note:* You may find that you want to use a method other than graphing to answer these questions. If so, use that method to *check* your answer *after* you have used the graph.

# Homework 12

# Who's Right?

| In | Out |
|----|-----|
| 3  | 8   |
| 5  | 12  |
| 9  | 20  |
| 15 | 32  |
| X  | ?   |

1. Andrew and Gladys were working on rules for In-Out tables. When they got to the table shown here, Andrew said that the *Out* at the bottom should be $2(X + 1)$. Gladys said it should be $2X + 2$. You need to decide who is right. Study the table and think about other In-Out pairs of numbers that you think would fit the pattern.

   If you think that either Andrew or Gladys is wrong, or that both are wrong, explain why you think so. If you think that they are both right, explain how there could be two different answers.

2. Find the area of each of the shapes shown here. That is, find out how many 1-foot–by–1-foot squares will fit in each without overlapping. Assume that all angles are right angles. *Do not assume* that these drawings are to scale.

   a.

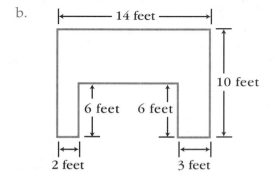

   b.

## Days 13–21

# *What's the Same?*

In everyday language, we can usually say the same thing in many different ways. This is also true about the language of algebra. Algebraic expressions that say the same thing are called "equivalent."

In mathematical work, you often need to be able to switch smoothly from one algebraic expression to an equivalent one. In the next section of this unit, you'll be looking at ways to do this and at how to use equivalent expressions to solve equations. You may be surprised to see that you can often use geometry to find equivalent algebraic expressions.

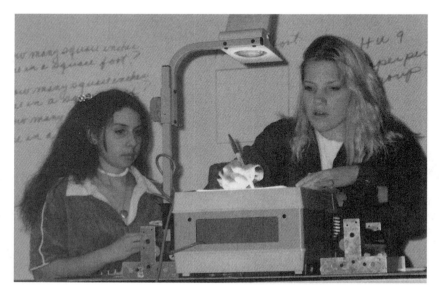

*Sofia Anis and Molly Berglund take delight in challenging the class to solve some of the equations they have scrambled.*

Interactive Mathematics Program

# A Lot of Changing Sides

A housing developer submitted plans to the city planner for some houses she wanted to build. The lots in the plan were all squares of the same size. But the city planner thought that this plan was boring and insisted that the developer introduce some variety. After some discussion, the planner and the developer decided that the lots should include other types of rectangles. So the developer proceeded to change the lengths of some of the sides of the lots.

For each of the changes that were made to the square lots, complete these tasks.

• Make and label a sketch of the lot, using the variable $X$ to represent the length of a side of the original square.

• Write an expression for the area of the new lot as a product of its length and width.

• Write an expression *without parentheses* for the area of the new lot as a sum of smaller areas. Use your sketch to explain this expression.

1. The original square lot was extended 4 meters in one direction and 3 meters in the other.

2. The original square lot was extended 5 meters in one direction only.

3. The original square lot was extended 10 meters in one direction and 9 meters in the other.

4. The original square lot was extended 1 meter in one direction and 25 meters in the other.

5. The original square lot was extended 2 meters in one direction and decreased 2 meters in the other.

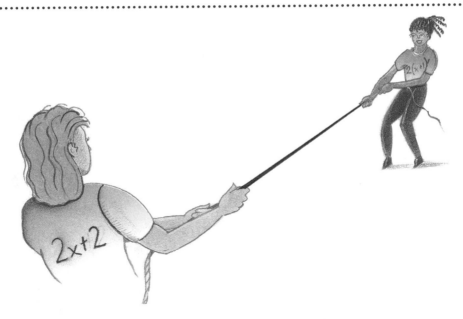

# Homework 13   Why Are They Equivalent?

You saw in *Homework 12: Who's Right?* that the two expressions $2(X + 1)$ and $2X + 2$ seem to give the same result no matter what number is substituted for $X$. In other words, the expressions appear to be equivalent. But it would be nice to be certain of this and to understand *why* the expressions are equivalent.

Randy, Sandy, and Dandy were having just that discussion. Read each of their explanations, and then do these four things.

1. Decide whether any, all, or just some of them are correct, and explain your decision.

2. State which explanation is the easiest for you to understand, and why.

3. State which explanation is most convincing to you, and why.

4. Adapt the explanation you understand best to explain in your own words why the expressions $3(X + 4)$ and $3X + 12$ are equivalent.

## *Randy's Explanation*

"We all know that $2A$ is twice $A$, which is $A + A$. Think of $2(X + 1)$ as being twice $X + 1$. In other words, it is equal to $(X + 1) + (X + 1)$. And $(X + 1) + (X + 1)$ is equal to $2X + 2$."

*Continued on next page*

# Sandy's Explanation

"It works with numbers! Check it out! If $X$ is 5, then $2(X + 1)$ is $2(5 + 1)$, which is the same as $2 \cdot 6$, which is 12. And, well, $2X + 2$ is $2 \cdot 5 + 2$, which is the same as $10 + 2$, which is also 12! Wow!"

# Dandy's Explanation

"Multiplication is how you find the area of a rectangle, you know, length times width. Basically, a product $ab$ can be thought of as the area of a rectangle with dimensions $a$ and $b$, like this:

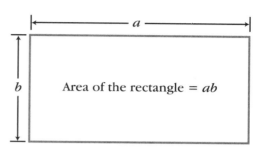

"The product $2(X + 1)$ can represent the area of a rectangle that is 2 units in one dimension and $X + 1$ units in the other. The length of $X + 1$ is like a segment of length $X$ next to a segment of length 1. The picture is something like this:

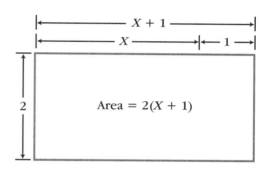

"A simple dividing line shows that this figure can be thought of as two rectangles, with areas $2X$ and 2, put together. Because we are talking about the same area, $2(X + 1)$ must equal $2X + 2$."

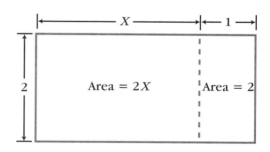

# Homework 14

# One Each Way

1. Find an equivalent expression without parentheses for each of these expressions.

   a. $5(A + 7)$

   b. $8(y - 4)$

   c. $2(W + 6)$

2. In *Homework 13: Why Are They Equivalent?* you saw three ways of thinking about why $2(X + 1)$ is equivalent to $2X + 2$. Now use those ideas to explain your work in Question 1.

   a. Use Randy's *repeated addition* method to explain why your answer to Question 1a is equivalent to $5(A + 7)$.

   b. Use Sandy's *numerical example* method to explain why your answer to Question 1b is equivalent to $8(y - 4)$.

   c. Use Dandy's *area model* method to explain why your answer to Question 1c is equivalent to $2(W + 6)$.

3. Find an equivalent expression without parentheses for each of these expressions. Use any method you like, but explain your work.

   a. $(r + 4)(r + 3)$

   b. $(3t + 1)(t + 5)$

# Distributing the Area

The figure on the right is a large rectangle made up of some smaller rectangles. The measurements of the smaller rectangles are shown using the variables *a*, *b*, *c*, and *d*.

Remember that you can compute the area of a rectangle by multiplying its length by its width.

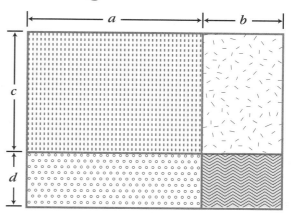

1. Use the "length times width" area formula and the variables *a*, *b*, *c*, and *d* to write an expression for the area of each of the smaller rectangles from the diagram.

   a. Area of the rectangle shaded like  = –?–

   b. Area of the rectangle shaded like ⬜ = –?–

   c. Area of the rectangle shaded like ⬜ = –?–

   d. Area of the rectangle shaded like ⬜ = –?–

*Continued on next page*

2. Next, look at certain combinations of rectangles and write each area in two ways.

- As the product of its length and width

- As the sum of two smaller areas

a.  Area of the figure shaded like  = –?–

b.  Area of the figure shaded like  = –?–

3. Write the area of the entire rectangle in *at least* two ways.

4. a.  Draw and label a rectangle whose area can be written as the product $(p + q + r)(x + y + z)$.

   b.  Show how to use your diagram to write the product $(p + q + r)(x + y + z)$ as an expression without parentheses.

5. *Challenge:* Draw a diagram that could be used as a model for finding an expression without parentheses that is equivalent to $(a + b)^3$.
   (*Hint:* We can't call it an *area* model.)

# Homework 15    The Distributive Property and Mystery Lots

## Part I: The Distributive Property

In its simplest form, the distributive property says that the expressions $a(x + y)$ and $ax + ay$ are equivalent. In other words, according to the distributive property, the equation

$$a(x + y) = ax + ay$$

is true no matter what numbers are substituted for $a$, $x$, and $y$.

Sometimes this property is used to replace an expression with parentheses by an equivalent expression without parentheses. For example, you can write $5(2x + 3)$ as $10x + 15$. This use of the distributive property is often called **multiplying through.** We also sometimes say that the factor 5 has been "distributed" across the sum $2x + 3$.

*Continued on next page*

1. Distribute the factor across each sum.

   a. $4(a + 9)$

   b. $3(5w + 2r)$

   c. $6t(2 + 3s)$

   d. $10c(u + v + w)$

The distributive property is also used in the reverse direction from the examples of Question 1. For instance, you can write $8x + 12y$ as $4(2x + 3y)$. This use of the distributive property is often called **taking out a common factor** (or just *factoring*). You might want to think of this as "undistributing."

2. Take out a common factor in each of these sums.

   a. $14d + 21e$

   b. $rg + rh$

   c. $2pq + 4pr$

   d. $6ab + 10ac$

## *Part II: Mystery Lots*

The developer from *A Lot of Changing Sides* has taken to writing plans for lot sizes in algebraic code. Unfortunately, the codes are giving the city planner a hard time.

3. Here is what the city planner found written in the developer's notes one day:

   "Build a lot whose area is $X^2 + 3X + 5X + 15$."

   Help the city planner by finding out what the developer planned to do with the original square lot. (Remember that the original lot had sides of length $X$.)

4. What do you suppose each of these two entries means?

   a. "Build a lot whose area is $X^2 + 4X + 6X + 24$."

   b. "Build a lot whose area is $X^2 + 6X + 2X + 12$."

5. Then the developer's entries got even more cryptic. Figure out for the city planner what each of these entries means.

   a. "Build a lot whose area is $X^2 + 9X + 18$."

   b. "Build a lot whose area is $X^2 + 7X + 10$."

   c. "Build a lot whose area is $X^2 + 6X + 8$."

   d. "Build a lot whose area is $X^2 + 5X + 4$."

# Homework 16    Views of the Distributive Property

The **distributive property** is an important general principle that can be used in many situations to write a mathematical expression in another form. Recall that in its simplest algebraic form, the distributive property can be expressed by an equation like this one.

$$a(x + y) = ax + ay$$

In words, you might state the distributive property this way.

> Multiplying a sum by something is the same as multiplying each term by that "something" and then adding the products.

In this assignment, you'll be looking at various ways to think about and use the distributive property.

## Multidigit Multiplication

You may not have realized that you've been using the distributive property every time you do multiplication that involves more than one-digit numbers. For example, the product 73 · 56 can be thought of as (70 + 3) · 56. Applying the distributive property, you would get 70 · 56 + 3 · 56.

You might write this problem in vertical form.

$$
\begin{array}{r}
56 \\
\times\, 73 \\
\hline
168 \\
3920 \\
\hline
\end{array}
$$

       168    (this is 3 · 56)

     3920    (this is 70 · 56)

*Comment:* People often omit the zero in 3920, simply multiplying 7 · 56 to get 392 and then writing 392 with the 2 lined up in the tens column.

*Continued on next page*

Each of the products 3 · 56 and 70 · 56 can also be found using the distributive property. To show all the details in the written multiplication, you might write it like this.

$$
\begin{array}{r}
56 \\
\times\,73 \\
\hline
18 \\
150 \\
420 \\
3500 \\
\hline
\end{array}
$$

       18    (this is 3 · 6)

   150    (this is 3 · 50)

   420    (this is 70 · 6)

3500    (this is 70 · 50)

Each of the numbers 18, 150, 420, and 3500 is called a **partial product.** Writing a multidigit multiplication showing all the partial products is sometimes called the **long form.**

In the usual written form of this problem, the partial products 18 and 150 are not shown individually. Instead, their sum, 168, is written. Similarly, we omit the partial products 420 and 3500 and simply write their sum, 3920. The numbers 168 and 3920, which are each the sum of two partial products, are sometimes referred to as **partial sums.**

1. Show how to find the product 32 · 94 using the long form, showing all the partial products.

## *Multiplication with a Diagram*

You can illustrate the product 73 · 56 with an area diagram like this one, in which each smaller rectangle represents one of the partial products. Notice that the areas of the two smaller rectangles on the right add up to the partial sum 168 while the areas of the two larger rectangles on the left add up to the partial sum 3920.

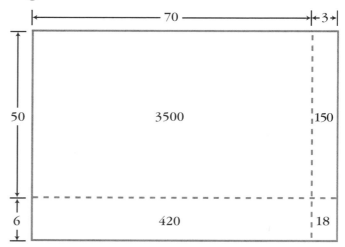

*Continued on next page*

2. Draw an area diagram like the one shown on the previous page to illustrate the product 32 · 94.

3. Show how to find the product 47 · 619 in two ways.

    a.  Using the long form

    b.  Using an area diagram

# Multiplying in Algebra Is Like Multiplying in Arithmetic

Multiplication of algebraic expressions can be done in a way that is similar to multidigit multiplication. For example, you can set up the problem $(x + 3)(2x + 5)$ in vertical form.

$$2x + 5$$
$$\underline{\times \ x + 3}$$

As with 73 · 56, this problem involves four separate products.

4. a.  Find this product using a vertical multiplication form. You can use either the long form or a shorter form.

    b.  Show how to do this problem using an area diagram.

# Prime Time

A **prime number** (also called simply a *prime*) is a whole number that has exactly two whole-number divisors: 1 and itself. For example, 7 is a prime number, because it has exactly two whole-number divisors: 1 and 7. On the other hand, 10 is not a prime, because it has four whole-number divisors: 1, 2, 5, and 10. A nonzero whole number with more than two whole-number divisors is called a **composite number.**

The number 1 is *not* considered a prime, because it has only one whole-number divisor, nor is it a composite number. Rather, it is considered a special case.

1. Examine each of the numbers from 2 through 30.

   a. Make a list of the numbers in this group that are primes.

   b. For each composite number from 2 through 30, write that number as a product of prime numbers. This is called the **prime factorization.** *Note:* You may need to use more than two factors, and you can use the same factor more than once. For example, the prime factorization of 12 is the product 2 · 2 · 3.

2. If an expression from Question 1b used a factor more than once, rewrite the expression using an exponent instead of repeating the factor. For example, write 12 as $2^2 \cdot 3$ instead of 2 · 2 · 3.

# POW 3          *Divisor Counting*

As you saw in *Prime Time,* a prime number is a whole number that has exactly two whole-number divisors. This POW is about counting the divisors for any whole number—not only primes. Throughout this problem, the word *divisor* will mean *whole-number* divisor.

The number 1 is a divisor of every whole number and every whole number is a divisor of itself. Therefore, every whole number greater than 1 has at least two distinct divisors and so must either be a prime or have more than two divisors. Your task in this POW is to figure out as much as you can about *how many* divisors a number has. You will probably find the concept of prime numbers useful both in conducting your investigation and in stating your conclusions.

*Continued on next page*

Here are some examples of questions to look at.

- What kinds of numbers have *exactly* three divisors? *exactly* four? and so on.

- Do bigger numbers necessarily have more divisors?

- Is there a way to figure out how many divisors 1,000,000 (one million) has without actually listing and counting them? How about 1,000,000,000 (one billion)?

- What's the smallest number that has 20 divisors?

But you should not answer only these questions. You should also come up with your own questions and look for generalizations.

# Write-up

1. *Subject of Exploration:* What were you exploring? What were your goals?

2. *Questions:* What questions did you ask yourself? Why did you ask them? Which ones did you decide to explore?

3. *Information Gathering:* Based on your notes, describe what you did to get data for your exploration.

   a. How did you get started?

   b. What approaches did you try?

   c. What information did you gather?

   d. When did you decide to stop, and why?

4. *Results and Conjectures:* What conjectures did you find as possible answers to your questions? What rules or patterns did you find in exploring your questions? If you can prove that your conjectures are right, do so. If you can explain why a particular rule or pattern works, do that as well. If possible, generalize your results.

5. *Evaluation*

6. *Self-assessment*

# Homework 17        Exactly Three or Four

1.  By definition, a number with exactly two whole-number divisors is a prime number. But what about numbers with exactly three whole-number divisors? Or exactly four?

    To get you started on *POW 3: Divisor Counting,* your first task in this assignment is to consider these two special cases.

    a.  Find several numbers that each have exactly three divisors and several others that each have exactly four divisors.

    b.  Examine your two lists and look for some explanations or patterns.

2.  Decide on at least one specific question about divisors that you want to try to answer as part of your POW, and state that question as clearly as you can.

# Taking Some Out, Part I

Do you remember the chefs from *Patterns* in Year 1? You used their situation to help with the arithmetic of positive and negative numbers. Well, thinking about temperatures can also be of help when finding equivalent expressions.

In each of the problems here, you should assume that when the action begins, the temperature of the cauldron is 0 degrees. As usual, every hot cube added to the cauldron increases the temperature by one degree and every hot cube removed from the cauldron lowers the temperature by one degree.

1. The chefs decided to put 50 hot cubes into the cauldron, but once they did so, they found that the cauldron was too hot. So two of the chefs reached in and removed some hot cubes. One chef removed a batch of 5 hot cubes and the other chef removed a batch of 10 hot cubes.

   a. What was the temperature when this was all done?

*Continued on next page*

   b. Write the entire process as a chef instruction in two ways:

     • With parentheses, showing the two batches of cubes being removed together

     • Without parentheses, showing the two batches of cubes being removed one batch at a time

2. Another time the chefs put 45 hot cubes into the cauldron and again found that the cauldron was too hot. Two chefs removed some hot cubes. One chef took out a batch of 8 hot cubes and the other took out a batch of 11 hot cubes.

   a. What was the temperature at the end of this process?

   b. As in Question 1b, write the entire process as a chef instruction in two ways.

3. The next time this happened, the chefs put in 60 hot cubes to begin with, and again two chefs took some out. The first chef removed a batch of 9 hot cubes, but the second chef forgot to count the number of cubes he removed.

   a. Create several rows for an In-Out table in which the *In* is the number of hot cubes the second chef removed and the *Out* is the final temperature.

   b. Find two rules for the table—one with parentheses and one without parentheses. Use *X* to represent the *In*.

   c. Graph your two rules on the graphing calculator and see if they give you the same graph.

# Homework 18     Subtracting Some Sums

1. Write each of these expressions as an equivalent expression without parentheses. Simplify your results where you can by combining like terms.

   a. $35 - (3a + 14)$

   b. $50 - (c + 17 + 2d)$

   c. $16 + 9s - (3s + 11)$

   d. $23 + 5w - 2(w + 7)$

2. The equations in the next series involve subtracting a sum that is in parentheses. Use whatever techniques make sense to you to solve these equations, but write an explanation of what you do.

   a. $54 - (t + 5) = 32$

   b. $29 - 2(x + 4) = 5$

   c. $6z + 17 - (2z + 5) = 56$

*Continued on next page*

3. Because the distributive property is so important in working with algebraic expressions, it's helpful to be able to apply it smoothly and with confidence. Find each of the products below, writing the results without parentheses and combining like terms when possible.

   a. $(x + 4)(x + 7)$

   b. $(2t + 3)(3t - 5)$

   c. $(4r - 3)(3r - 2)$

   d. $(x^2 + 3x + 2)(x + 6)$

4. Soon you'll be combining ideas about equivalent expressions with your insights learned from work with the mystery bags in order to solve more complex equations. In preparation for that, here are some mystery bag problems for you. Solve these equations to find out how much gold is in each mystery bag.

   a. $26M + 37 = 19M + 58$

   b. $46a + 95 = 83a + 29$

   c. $153x + 149 = 327x + 73$

# *Taking Some Out, Part II*

The chefs are continuing to play around with their cauldron. Use each of these problems to investigate ways to write different expressions for the same situation. As in *Taking Some Out, Part I,* each situation begins with a temperature of 0 degrees.

1. The chefs tossed 75 hot cubes into the cauldron. A few minutes later, one of the chefs reached in to remove some of them. She already had 12 hot cubes in her hands when she stumbled and 4 of those hot cubes fell back in.

   a. What was the temperature at the end of this process?

   b. Using the numbers in the problem, write the entire process in two ways.

   • Show the initial amount put in and subtract an expression in parentheses to show what was removed altogether.

   • Show the initial amount put in, use subtraction to show the whole batch of cubes being removed, and use addition to show some of that batch going back in. Your expression should not have parentheses.

*Continued on next page*

2. The next time, 62 hot cubes were put in originally. A chef then removed 14 of them, but 9 of the 14 fell back into the cauldron.

   a. What was the temperature at the end of this process?

   b. As in Question 1b, write the entire process in two ways.

3. The third time, 54 hot cubes initially were tossed in. A chef reached in and grabbed 25 of them, but first 6 of the 25 and then 4 more fell back in.

   a. What was the temperature at the end of this process?

   b. Write this entire process in *at least three* different ways.

4. Once again, the chefs tossed a big batch of hot cubes into the cauldron. Someone reached into the cauldron and pulled a handful of them out, but part of that handful fell back in. Write two different general expressions, one with parentheses and one without, describing what happened. Use different variables to represent the initial amount put in, the amount initially removed, and the amount that fell back in.

# Homework 19

# Randy, Sandy, and Dandy Return

## *Part I: Generalizing the Distributive Property*

Randy, Sandy, and Dandy are having another of their heated arguments. This time they aren't discussing why the distributive property is true, but are trying to find other principles that might be based on similar reasoning.

1. Randy says, "I use the distributive property all the time, even when it just involves multiplication." In other words, Randy thinks that $a(bc) = (ab) \cdot (ac)$.

   Is she correct? Try substituting some numbers to find out. If she's right, explain why. If she's wrong, rewrite the right side of the equation to make her statement correct.

2. Sandy then says, "I use the distributive property all the time, too." (This makes Randy worry a little about what she thinks.) "And I use it with additions all over the place." What Sandy thinks is that $a + (b + c) = (a + b) + (a + c)$.

   Is she correct? Try substituting some numbers to find out. If she's right, explain why. If she's wrong, rewrite the right side of the equation to make her statement correct.

3. Dandy thinks they are both confused and says, "I don't know what you two are thinking of, but I know that $a - (b - c) = a - b + c$."

   Is he correct? Try substituting some numbers to find out. If he's right, explain why. If he's wrong, rewrite the right side of the equation to make his statement correct.

## *Part II: Distributing Mystery Bags*

4. For each of the equations shown here, first use the distributive property (correctly!) to remove the parentheses on each side of the equation, and then combine terms on each side and solve the resulting equation.

   a. $4(M + 2) + 7 = 6(M + 1) + 2$

   b. $3(x + 9) + 2(3x + 4) = 7(x + 11)$

   c. $5(2x - 3) + 3(x + 8) = 4(x + 6) + 3(x - 4)$

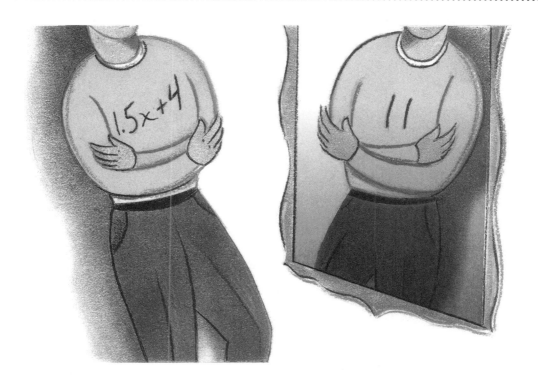

# Homework 20        Equation Time

Equations play an important role in mathematics, and the concept of equivalent equations is a valuable tool in solving equations. These problems give you a chance to apply some of what you know about equivalent equations.

1. A student trying to solve the equation $1.5x + 4 = 11$ wrote what is shown below. Is this correct? If not, why not?

$$1.5x + 4 = 11$$

$$3x + 4 = 22 \quad \text{(multiplying both sides by 2)}$$

$$3x = 18 \quad \text{(subtracting 4 from both sides)}$$

$$x = 6 \quad \text{(dividing both sides by 3)}$$

2. In your work with similar triangles, you have used a particular type of equation called a *proportion,* which is a statement that two ratios are equal.

*Continued on next page*

For example, if you know that the two quadrilaterals shown here are similar, you might come up with the proportion

$$\frac{x}{5} = \frac{x+2}{8}$$

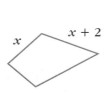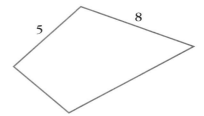

One of the principles for getting equivalent equations is that you can multiply both sides of the equation by the same thing.

a. Apply this principle to the equation $\frac{x}{5} = \frac{x+2}{8}$, first multiplying both sides by 5 and then multiplying both sides of the result by 8 (or simply multiply both sides by 40).

b. Simplify and solve the equation you got in Question 2a and check your solution in the original equation.

3. Solve each of these equations, explaining each step clearly.

a. $41 + 7d - 5(d + 7) = 8d + 1$

b. $8w - 3(2w - 9) = 7(w + 2)$

# Scrambling Equations

Usually, the concept of equivalent equations is used to make things simpler. But in this activity, you're going to make things more complicated. For example, look at the sequence of equations shown below.

$$x = 1$$

$$6x = 6$$

$$6x - 3 = 3$$

$$\frac{6x - 3}{2} = 1.5$$

All of these equations are equivalent, because they all have the same solution. You should be able to see what was done to each equation to get the one below it.

In this activity, you will begin by writing down a *very simple* equation (like $x = 1$). Then you'll write down an equivalent equation that's more complicated, and then something equivalent to that, and so on.

*Continued on next page*

This activity has some very precise rules. You will be changing your equation exactly three times. At each stage, you can do any one of these four things.

- You can add the same integer to both sides of the equation.

- You can subtract the same integer from both sides of the equation.

- You can multiply both sides of the equation by the same nonzero integer.

- You can divide both sides of the equation by the same nonzero integer.

Remember that you are to do *exactly three* of these steps (in any order). For instance, the example shown above uses multiplication, then subtraction, and then division. At any time in the process, you're also allowed to do arithmetic steps to simplify the right side of the equation.

When you are done with this process, copy your final, complicated equation onto one side of a sheet of paper and put your original equation on the reverse side. This sheet will be exchanged with another group, and you will have the opportunity to "uncomplicate" someone else's scrambled equation.

# Homework 21   More Scrambled Equations and Mystery Bags

## Part I: More Scrambled Equations

This assignment involves the same steps for getting equivalent equations that were described in *Scrambling Equations*.

1. The equations here show one sequence of three steps to "scramble" the equation $x = 3$.

$$x = 3$$
$$x - 5 = -2$$
$$10(x - 5) = -20$$
$$\frac{10(x - 5)}{4} = -5$$

   a. Describe what was done at each step.

   b. Check that $x = 3$ is a solution to the final equation in the sequence, and show your work.

*Continued on next page*

For Questions 2 through 4, do two things

- "Uncomplicate" each equation until you get back to a simple equation of the form "$x$ = some number."

- Take the value of $x$ you get from the simple equation and substitute it back into the original equation in order to check that it makes the "complicated" equation true.

2. $3x - 5 = -2$

3. $\dfrac{x - 6}{4} + 1 = 7$

4. $4\left(\dfrac{x}{3} + 6\right) - 8 = 20$

# Part II: More Mystery Bags

Earlier in this unit, you used the idea of a pan balance to solve mystery-bag problems.

The equations here might come from such problems. Solve them using the concept of equivalent equations, but also think about how each step you do is related to the pan-balance model.

5. $11t + 13 = 7t + 41$

6. $12 + 7w = 4w + 21$

7. $8(x + 3) + 19 = 15 + 2(x + 35)$

**Days 22–26**

# The Linear World

A straight line is one of the simplest types of graphs. In *Homework 11: Line It Up,* you saw that certain algebraic expressions lead to graphs that are straight lines.

Linear equations and linear functions are important in many applications of mathematics, and you already know a great deal about them. In an activity called *Get It Straight,* you'll work with your classmates to learn even more.

*Terria Galvez and Mekea Harvey continue their work on linear functions and straight-line graphs.*

# Old Friends and New Friends

## Part I: Old Friends

Over the course of this unit, you have set up equations for a variety of problems. The examples here give brief reminders of some of those problems. With each problem is an equation that might have been used to help solve the problem.

Your task in this assignment is to solve these equations. Although you could probably solve them by trial and error or by graphing, you are to solve them here using equivalent equations. Show the steps you use to get from the equations to the solutions, and check your answers by substituting them back into the original equations. *Note:* Because these problems are stated here without details, you do not need to explain how each equation fits its problem.

1. Problem: Find the payoff that would give Al a total gain of 25 points (Question 3 from *Memories of Yesteryear*).

   Equation: $25x - 75 \cdot 2 = 25$

*Continued on next page*

2. Problem: Find the distance around the lake (Question 2 from *Homework 4: Running on the Overland Trail*).

   Equation: $2d + 4 = 10$

3. Problem: Find the number of days the Sawyers would need to catch up (Question 3 from *Catching Up*).

   Equation: $30N = 24N + 120$

4. Problem: Find the length of Nelson's shadow (Question 2 from *Lamppost Shadows*).

   Equation: $\dfrac{S}{S + 20} = \dfrac{6}{25}$

   (*Note:* This isn't exactly in the form of a linear equation, but it is essentially equivalent to a linear equation. You might look at Question 2 of *Homework 20: Equation Time* for ideas.)

## Part II: New Friends

These equations don't come from specific problems, but that doesn't affect the algebra. Solve each of them using equivalent equations, and show the steps you use.

5. $4(t + 5) + 3 = 7t + 19$

6. $6W - (2W + 1) = 3(W - 10)$

# Homework 22

# New Friends Visit Your Home

## Part I: Solve It!

As you've seen, linear equations come up in many situations. Sometimes the equations are simple, and sometimes they are complicated. Use the method of equivalent equations to solve each example here, and check your solutions by substituting into the original equations.

1. $7t - 5 = 10t + 8 - 4t$

2. $6(x - 2) = 4(x + 3) - 32$

3. $8r + 1 = 12r + 27$

4. $5(w + 4) - 3(w + 2) = 3(w + 3) - (w - 5)$

5. $6g - (3g + 8) = 16$

6. $7 - 4d = 3d - 9d + 25$

7. $6 + 4(y + 2) = 10 - 4y$

## Part II: Write It!

Make up a situation for which the equation $5 + x = 21 - x$ might be appropriate. Identify what the variable represents in the situation you create.

# Homework 23   From One Variable to Two

You've seen that a linear equation in one variable such as $3x + 4 = 2(x - 1)$ has a unique solution. The equation $x + 2y = 5 + 3x + y$ is also a linear equation, but it includes two variables and has more than one solution. This assignment looks at what it means to "solve" an equation like this.

1. Begin with the simpler two-variable equation, $y = 2x + 3$.

   a. Find at least three number pairs that fit this equation.

   b. Plot the number pairs you found in part a.

2. Now use the equation $x + 2y = 5 + 3x + y$.

   a. Find at least three number pairs that fit this equation. *Hint:* Pick a number for one of the variables and then find a value for the other variable that fits the equation.

   b. Plot the number pairs you found in part a.

3. Write an equation that is equivalent to $x + 2y = 5 + 3x + y$ but that expresses $y$ in terms of $x$. In other words, your equation should have $y$ by itself on the left and an expression involving $x$ on the right. This is called **solving for $y$ in terms of $x$.** *Suggestion:* Think of the equation as a mystery-bag problem in which $x$ and $y$ represent the weights for two different-size mystery bags.

# Get It Straight

You know that any equation involving $x$ and $y$ can be used to create a graph. The graph is defined as the set of all those points whose coordinates fit the equation.

Some equations have graphs that are straight lines. These are called **linear equations.** When a linear equation expresses $y$ in terms of $x$, it can be referred to as a **linear function.** All linear functions can be simplified so that they fit the form $y = ax + b$, where $a$ and $b$ represent two numbers.

The number $a$ is referred to as the **coefficient of** $x$ and the number $b$ is called the **constant term.** Keep in mind that these numbers can be positive, negative, or zero, and they can also be identical.

In this activity, you will investigate linear functions and straight-line graphs.

Here are some questions to explore.

- How do you change the equation in order to change the "slant" of its graph?

- How do you change the equation in order to shift the whole graph up or down?

*Continued on next page*

- When do two linear functions give parallel lines (lines that never meet)? Why?

- What linear functions give horizontal lines? Why?

- When do two linear functions give lines that are mirror images of each other with the *y*-axis as the mirror? Why?

- When do two linear functions give perpendicular lines (lines that form a right angle)? Why?

Do not feel limited by these questions—let your imagination soar! Keep track of any other interesting questions you think of, even if you can't answer them.

# *Write-up*

You should do a written report of what you learn, using these categories.

1. *Questions:* What questions did you ask yourself? Why did you ask them? Which ones did you decide to explore?

2. *Results and Conjectures:* What conjectures did you find as possible answers to your questions? What rules or patterns did you find in exploring your questions? If you can prove that your conjectures are right, do so. If you can explain why some rule or pattern works, do that as well. If possible, generalize your results.

# Homework 24     A Distributive Summary

The distributive property played an important role in this unit. This assignment on that important idea will be part of your unit portfolio.

Write as complete an explanation of the distributive property as you can. You should touch on at least these issues.

- What it says

- Why it's true

- Examples of situations in which you would use it

Interactive Mathematics Program

# Homework 25        All by Itself

You've seen that a linear equation in two variables can be transformed into a linear function by writing one of the variables in terms of the other variable. This assignment continues that theme.

1. In the activity *Fair Share on Chores* (from the Year 1 unit *The Overland Trail*), three boys and two girls were responsible for watching the animals for a total of ten hours, with the boys and the girls each having a shift of a certain length. (*Remember:* Families considered this fair in light of other chores the boys and girls had to do.)

    If $B$ represents the length of each boy's shift and $G$ represents the length of each girl's shift, then the fact that the total time is ten hours can be represented by the equation

    $$3B + 2G = 10$$

*Continued on next page*

Interactive Mathematics Program        75

Solve this equation for $B$ in terms of $G$. In other words, find an equivalent equation of the form

$$B = \text{an expression involving } G$$

*Hint:* One approach is to imagine that you knew the length of each girl's shift (the value of $G$) and to think about how you would figure out the length of each boy's shift (the value of $B$).

2. In Question 1, you worked with a linear equation that came from the context of a real-life situation. In these examples, all you have are the equations. In each case, solve the equation for the specified variable.

   a. Solve this equation for $v$ in terms of $w$.

   $$2v + 7 = w - 3$$

   b. Solve this equation for $s$ in terms of $r$.

   $$4r + 5s + 2 = 8r - s + 7$$

   c. Solve this equation for $y$ in terms of $x$.

   $$5y - 2x + 1 = 3(y + x) - (x - 5)$$

   d. Solve the equation from Question 2b for $r$ in terms of $s$.

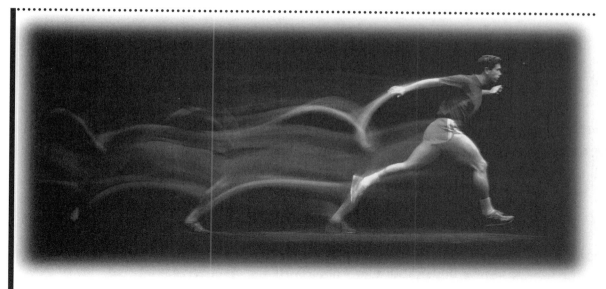

# Homework 26    More Variable Solutions

In *Homework 25: All by Itself,* you looked at some linear equations and found equivalent equations expressing one variable in terms of another. In this assignment, only Question 1 involves a linear equation.

1. Solve this linear equation for $z$ in terms of $x$.

$$3(x - z) - 2(5 - x) = 4z + 2 - 6(x + z)$$

On the remaining examples, you may find it useful to imagine that certain variables represent numbers or actually to replace them with numbers. Then think about how you would solve the given equation for the remaining variable. This approach is used in Question 2.

2. The kinetic energy of a moving object is given by the formula

$$W = \frac{1}{2} mv^2$$

where $W$ represents the kinetic energy, $m$ represents the mass of the object, and $v$ is the object's velocity.

   a. Suppose $W = 30$ and $v = 3$. What is the numerical value of $m$? (Don't worry about the units involved for energy, mass, or velocity.)

   b. Use your work from Question 2a to solve the equation for $m$ in terms of $W$ and $v$.

*Continued on next page*

3. Solve the equation from Question 2 for $v$ in terms of $W$ and $m$.

4. Coulomb's law states that the force of attraction or repulsion between two electrical charges is proportional to the product of their charges and inversely proportional to the square of the distance between them. In symbols, Coulomb's law can be expressed by the equation

$$F = \frac{kq_1q_2}{r^2}$$

where $F$ is the force, $q_1$ and $q_2$ are the charges, and $r$ is the distance between them. The letter $k$ represents a number called a **constant of proportionality.**

a. Solve this equation for $q_1$ in terms of the other variables.

b. Solve this equation for $r$ in terms of the other variables.

*Note:* The small numbers 1 and 2 in the notation $q_1$ and $q_2$ are called **subscripts.** Subscripts are often used when several variables represent similar things in a problem situation. In this case, there are two particles with electric charges, and the sizes of the charges are each represented using a single variable consisting of the letter $q$ and a subscript number.

**Days
27–31**

# Beyond Linearity

So now you can solve any linear equation! But what about equations that aren't linear? Are there any systematic ways to solve them?

In the final segment of this unit, you'll see how useful graphs and graphing calculators can be in solving more complicated equations.

*Kasey Kure writes his thoughts in response to an IMP problem.*

# *Where's Speedy?*

Speedy is the star runner for her country's track team. Among other things, she runs the last 400 meters of the 1600-meter relay race.

A sports analyst recently studied the film of a race in which she competed. The analyst came up with this formula to describe the distance Speedy had run at a given time in the race.

$$m(t) = 0.1t^2 + 3t$$

In this formula, $m(t)$ gives the number of meters Speedy had run after $t$ seconds of the race, with both time and distance measured from the beginning of her 400-meter segment of the race. (This formula might not be very accurate, but you are to work on this activity as if it were completely correct.)

| $t$ | $m(t)$ |
|-----|--------|
|     |        |

1. Use the formula for $m(t)$ to fill in several rows of this In-Out table to show how far Speedy had run at different times of the race.

2. Use the table from Question 1 to make a graph that represents this situation. You may need to add more information to your table in order to obtain a graph that shows the entire time she is running.

3. Write an equation using the variable $t$ that you could solve to get the answer to the question "How long does it take Speedy to run her first 200 meters?"

4. Graph the function $m(t) = 0.1t^2 + 3t$ on the graphing calculator and use your graph to get an approximate solution to your equation from Question 3.

# Homework 27

# A Mixed Bag

1. Find the numerical solution to each of these linear equations. Be sure to check your answers.

   a. $5a + 3 = 7a - 9$

   b. $4(t + 7) - 2(t - 3) = 8t + 11$

2. Solve each of these equations for the variable indicated in terms of the other variables.

   a. Solve for $r$ in terms of $u$.

   $$8u + 5r - 14 = 0$$

   b. Solve for $y$ in terms of $a$, $b$, $c$, and $x$.

   $$ax + by = c$$

   c. Solve for $m$ in terms of $n$.

   $$3m^2 + 5n^2 = 21$$

# To the Rescue

A helicopter is flying to drop a supply bundle to a group of firefighters who are behind the fire lines. At the moment when the helicopter crew makes the drop, the helicopter is hovering 400 feet above the ground.

The principles of physics that describe the behavior of falling objects state that when an object is falling freely, it goes faster and faster as it falls. In fact, these principles provide a specific formula describing the object's fall, which can be expressed this way:

Suppose that the object's height off the ground when it begins to fall, at time $t = 0$, is $N$ feet, and use $h(t)$ to represent the object's height off the ground $t$ seconds after being dropped. Then the function $h(t)$ is given by the equation $h(t) = N - 16t^2$.

So in the case of the falling supplies, the formula is $h(t) = 400 - 16t^2$, because the supply bundle is 400 feet off the ground when it starts to fall.

1. How many seconds will it take the bundle to reach the ground? (*Hint:* What is $h(t)$ when the bundle reaches the ground?)

2. Write an equation that you could use to find out how many seconds it takes until the supply bundle is 100 feet off the ground.

3. Use a graphing calculator to find an approximate solution to your equation from Question 2.

4. Explain how you could check your solution from Question 2 using the formula $h(t) = 400 - 16t^2$.

# Homework 28

# Swinging Pendulum

In Question 4 of *Memories of Yesteryear*, you read about a group of students doing experiments with pendulums. They expressed the period of a pendulum as a function of its length by the formula $P = 0.32\sqrt{L}$, where $P$ is the time of one period (expressed in seconds) and $L$ is the length of the pendulum (expressed in inches).

1. Use the students' formula to find the period for pendulums with each of these lengths.

   a. 4 inches

   b. 16 inches

   c. 50 inches

   d. 10 feet

2. Use your answers from Question 1 and other data that you obtain from the formula to sketch a graph of the function. Use scales for your axes that are appropriate for your answers to Question 1.

After they found their formula, the students decided to build a clock using a pendulum. They wanted the period of the pendulum to be exactly one second.

3. Write an equation that they could use to find the correct length for their pendulum.

4. Find a solution to your equation. (Give an approximate value if necessary.)

5. Explain in words how you could use a graph of the function to solve the equation.

---

# Mystery Graph

The graph below shows the variable $y$ as a function of $x$, but it doesn't give a formula for this function. Instead, the graph is labeled with the generic function equation, $y = f(x)$.

Answer these questions based on the graph. Give approximate answers if necessary and state any assumptions you make about any portion of the graph that isn't visible.

1.  a.  Find $f(4)$. That is, what number would you get for $y$ if you substituted 4 for $x$?

    b.  Find $f(0)$.

    c.  Find $f(-1)$.

    d.  Find $f(-4)$.

2.  Find all solutions to the equation $f(x) = 0$. That is, find all the values of $x$ for which $y$ is 0.

3.  Solve each of these equations, giving all possible solutions.

    a.  $f(x) = 7$

    b.  $f(x) = 1$

    c.  $f(x) = -2$

    d.  $f(x) = -5$

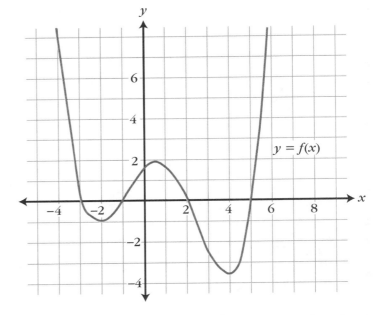

4.  a.  Find the maximum point for the part of the function between $x = -3$ and $x = 3$. That is, what point with an $x$-coordinate between $-3$ and 3 has the largest $y$-coordinate?

    b.  Find the minimum point for the part of the function between $x = -3$ and $x = 3$.

5.  Solve the inequality $f(x) > 0$. That is, find the values for $x$ that give a positive value for $y$. Describe all possible answers.

# Homework 29　　　　Functioning in the Math World

## *Part I: Expressions, Graphs, Tables, and Situations*

Algebraic expressions, graphs, and In-Out tables are three ways of representing the concept that mathematicians refer to using the word **function.** Mathematicians often blur the distinctions among these three representations, referring to them all as "the function."

Functions also are often connected to real-life situations. In this assignment, you will look at how these four ideas—expressions, graphs, tables, and situations—are related to one another. (Sometimes people use the phrase *rule of four* to refer to these four ways of thinking about functions.)

1.  The area of a square is determined by the length of any of its sides. For instance, if the length of a side is 7 inches, then the area is 49 square inches. Therefore, we can say that the area is *a function of* this length.

    a.  Make an In-Out table to go with this situation and fill in several rows for this table. (The values 7 and 49 would make up one row of your table.)

*Continued on next page*

b. Express the relationship between area and length in terms of an equation, explaining any variables you use.

c. Make a graph of your equation.

2. The equation $y = 3x + 1$ can be used to define a function.

a. Make an In-Out table to go with this equation and fill in several rows for this table.

b. Sketch the graph of the equation.

c. Create a situation for which this equation might be appropriate. Explain the role of the variables $x$ and $y$ in your situation.

# Part II: Another Mystery Graph

3. Study the graph of the function *h* shown below and then answer the questions. (You should consider only the part of the function shown in this graph.)

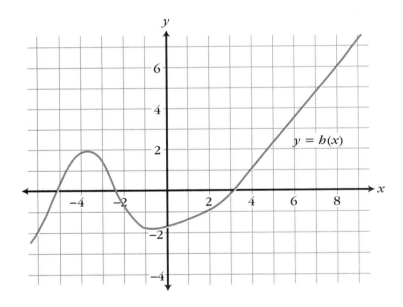

a. Estimate the value of $h(5)$.

b. Estimate the *y*-intercept of the function *h*.

*Continued on next page*

c.  Estimate the *x*-intercepts of the function *h*.

d.  Estimate all solutions to the equation $h(x) = -1$.

# Part III: Making Graphs That Fit Conditions

In these problems, you are asked to sketch graphs that fit certain conditions. Many graphs will work, but you need to create only one of them. *Note:* Most graphs are infinite, so the graph you actually draw will probably be only part of the total graph.

4. Sketch the graph for a function *f* that satisfies both of these conditions.

    • The function *f* has exactly three *x*-intercepts.

    • $f(2) = 5$

5. Sketch the graph for a function *k* that satisfies all of these conditions.

    • The function *k* has no *x*-intercepts.

    • $k(-1) = 2$

    • $k(3) = 1$

    • $k(5) = 6$

# *The Graphing Calculator Solver*

In *Where's Speedy?* the expression $0.1t^2 + 3t$ told you how far Speedy would run in the first $t$ seconds. The expression was used to define the function $m(t)$.

You then wrote and solved equations with this expression to answer questions about Speedy. For example, you solved the equation $0.1t^2 + 3t = 200$ to find out how long it would take Speedy to run the first 200 meters of her part of the race. You may have used the graph of the function $m(t)$ to solve such equations.

*Continued on next page*

# Equations, Functions, and Graphs

You can use this method to solve equations even when the equation doesn't come from a real-life situation. For example, even if you saw the equation $0.1t^2 + 3t = 200$ out of context, you could still enter the expression $0.1X^2 + 3X$ into a graphing calculator to define a function and get a graph. You could then solve the equation (approximately) by using the trace feature to find the values of $X$ that make $Y$ equal to 200. (In addition to the solution you found earlier, there is also a negative solution for $X$, which you would ignore if $X$ represented time.)

# A Graph Is Like an Answer Key

There's nothing magical about the expression $0.1t^2 + 3t$. Any meaningful expression that can be entered into a graphing calculator can be used to define a function. Once you've entered the expression, the graph of the function becomes an answer key for an entire family of equations involving that expression. For example, the graph of Speedy's function doesn't just help you solve the equation $0.1t^2 + 3t = 200$. It also helps you solve $0.1t^2 + 3t = 100$, $0.1t^2 + 3t = 63$, or any similar equation.

# Your Task

Your task in this activity is to use the graphing calculator to solve the equations given below. Give your answers to the nearest tenth.

*Note:* For simplicity, these equations have been chosen so that their solutions all lie between $x = -5$ and $x = 5$.

1. $2x^2 + 5x + 7 = 20$

2. $x^3 + 4x^2 - 5x + 1 = 12$

3. $x^4 - x^3 + 3x^2 + 5x = 6$

4. $x^3 + 4x = 2x^2 + 7x - 1$ (*Hint:* Use the expressions on each side of the equation to define two different functions. Then think of the original equation as asking for the value of $x$ that gives the same function value for both functions.)

# Homework 30     A Solving Sampler

You have used several ideas and methods for solving equations as part of this unit.

- Guess and check

- The mystery-bags model

- "Unscrambling" equations (equivalent equations)

- Graphing

In this activity, you will examine each of these methods as part of the preparation for your *Solve It!* portfolio.

1. Begin with the guess-and-check method.

   a. Summarize how the method works.

   b. Select an activity from the unit in which that method played an important role and attach that activity.

   c. Make up an equation for which you would use that method.

2. Do parts a, b, and c of Question 1 for the mystery-bags model.

3. Do parts a, b, and c of Question 1 for the equivalent-equations method.

4. Do parts a, b, and c of Question 1 for the graphing method.

# "Solve It!" Portfolio

Now that *Solve It!* is completed, it is time to put together your portfolio for the unit. This activity has three parts.

- Writing a cover letter summarizing the unit

- Choosing papers to include from your work in this unit

- Comparing this unit to Year 1 IMP units and to traditional algebra

## Cover Letter for "Solve It!"

Look back over *Solve It!* and describe the main mathematical ideas of the unit. This description should give an overview of how the key ideas were developed. In compiling your portfolio, you will be selecting some activities that you think were important in developing the key ideas of this unit. Your cover letter should include an explanation of why you selected particular items.

*Continued on next page*

# *Papers from "Solve It!"*

Your portfolio for *Solve It!* should contain these items.

- *Homework 30:A Solving Sampler.*
  Include both what you wrote about the different methods for solving equations
  and the sample activities you chose.

- *Get It Straight.*
  Include your write-up of your work on this activity.

- *Homework 24:A Distributive Summary.*

- A Problem of the Week.
  Select one of the three POWs you completed during this unit (*A Digital Proof*
  or *Tying the Knots* or *Divisor Counting*).

- Other high-quality work.
  Select one or two other pieces of work that represent your best efforts. (These
  can be any work from the unit—Problem of the Week, homework, classwork,
  presentation, and so on.)

# *"Solve It!" and Algebra*

This unit is more traditional than most of the IMP units. For example, it doesn't have a
central problem, and it involves manipulating algebra symbols. Discuss your reaction to
this type of unit. You might comment on these issues.

- How did you like this unit compared to Year 1 units with a central problem?

- Are you glad you did a unit emphasizing these traditional skills? Why or why not?

- How does the material in this unit compare with your idea of what algebra is?

# Appendix

# *Supplemental Problems*

As in Year 1, the supplemental problems for each unit pursue some of the themes and ideas that are important in that unit. Some of the supplemental problems in *Solve It!* continue the theme of looking back at ideas from Year 1. Others pursue ideas about equivalent expressions or follow up on ideas from the POWs. Here are some examples.

• *What to Expect* and *Carlos and Betty* give you more opportunities to work with the concept of expected value from the Year 1 unit *The Game of Pig.*

• *Same Expectations* and *Preserve the Distributive Property* give you a chance to extend your understanding of the distributive property and how it is used.

• *Ten Missing Digits* and *The Locker Problem* continue themes from *POW 1: A Digital Proof* and *POW 3: Divisor Counting.*

# *What to Expect?*

One of the problems from *Memories of Yesteryear* involved Al and Betty and the spinner shown below. This problem poses some more questions about that spinner.

1. a. If Al wins 4 points when the spinner lands on the gray area, what is his expected gain or loss per spin in the long run?

   b. If Al wins 10 points when the spinner lands on gray, what is his expected gain or loss per spin in the long run?

2. Al likes playing spinner games. He's willing to lose an average of $\frac{1}{4}$ point per spin in order to keep playing. Assume that Betty still wins 2 points when the spinner lands on white. What payoff should Al be willing to take each time the spinner lands on gray so that his expected loss is $\frac{1}{4}$ point per spin?

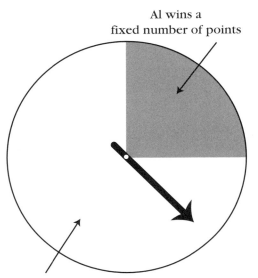

Al wins a fixed number of points

Betty wins 2 points

3. Make an In-Out table, based on the spinner shown above, in which the *In* is the amount Al wins when the spinner lands on gray and the *Out* is Al's average gain or loss per spin for that payoff. (Betty always gets 2 points when she wins.)

4. Find a rule for your In-Out table.

# Carlos and Betty

Carlos also likes spinner games. He has a spinner like the one at the right, in which Betty wins when the spinner lands in the white area ($\frac{2}{3}$ of the time). She gets 2 points from Carlos every time she wins.

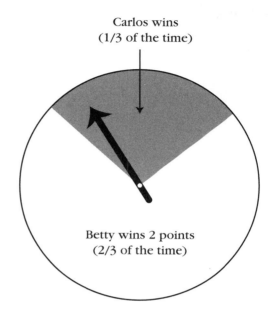

Carlos wins
(1/3 of the time)

Betty wins 2 points
(2/3 of the time)

1. How many points should Carlos win when the spinner lands in the gray area in order for the game to be fair? Explain how you got your answer.

2. Betty thinks it's fun to play spinner games. In fact, she is willing to lose points in the long run in order to keep Carlos interested in playing.

How many points should she give him for the spins he wins if she wants his average gain in the long run to be

a. $\frac{1}{10}$ point per spin?

b. $\frac{1}{2}$ point per spin?

Explain how you got your answers.

*Hint on both Question 1 and Question 2:* Experiment with various payoff amounts, find Carlos' expected value for each, and make an In-Out chart.

# *Ten Missing Digits*

In *Is It a Digit?* you had to fill in five empty boxes, labeled 0 through 4, in a way that satisfied certain conditions. In this problem, you have to solve a harder version of that problem. Specifically, consider the ten boxes below:

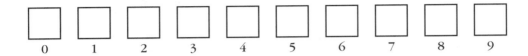

0    1    2    3    4    5    6    7    8    9

The rules are that you must put a digit from 0 to 9 in each of the boxes so that these conditions hold.

- The digit you put in the box labeled "0" must be the same as the number of 0's you use.

- The digit you put in the box labeled "1" must be the same as the number of 1's you use.

- The digit you put in the box labeled "2" must be the same as the number of 2's you use, and so on.

As in *Is It a Digit?* you are allowed to use the same digit more than once.

There may be more than one solution to this problem, so part of your task is to show that you have all the possible answers.

# Same Expectations

You may have noticed in your work with expected value in *The Game of Pig* that it didn't matter how many games or how many spins you used as "the long run." Here's your chance to see why.

The spinner shown here is the same as in Question 3 of *Memories of Yesteryear,* except that now it shows Al winning 5 points each time the spinner lands on gray. As before, Betty wins 2 points each time the spinner lands on white.

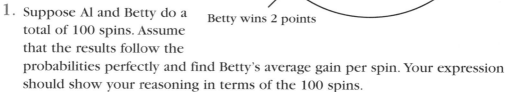

Al wins 5 points

Betty wins 2 points

1. Suppose Al and Betty do a total of 100 spins. Assume that the results follow the probabilities perfectly and find Betty's average gain per spin. Your expression should show your reasoning in terms of the 100 spins.

2. Repeat Question 1 but use 1000 spins this time.

3. Repeat Question 1 but this time use the variable *N* for the number of spins. Use a little algebra to prove that if the results follow the probabilities perfectly, then Betty's average gain per spin is the same no matter how many spins there are.

# *Preserve the Distributive Property*

Some of the rules for multiplication with integers make intuitive sense, but others can be confusing. For example, the fact that 3(–2) = –6 can be explained in terms of repeated addition:

$$3(-2) = -2 + (-2) + (-2)$$

The fact that –2 + (–2) + (–2) = –6 also seems reasonable, so we have 3(–2) = –6.

It also seems intuitively reasonable to most people that 0(–2) = 0. But many people have a hard time understanding why the product of two negative numbers should be positive. It turns out that the distributive property can provide an explanation for this fact.

Your first task in this activity is to use the expression (–3 + 3)(–2) to show that (–3)(–2) = 6. The key idea is to evaluate (–3 + 3)(–2) in two different ways, using the distributive property in one of the ways. Then generalize this example to explain why the product of any two negative numbers must be positive. You may assume that the product of a positive number and a negative number is negative.

# The Locker Problem

Louise is walking through the hallway of her school past the row of lockers on the first day of school. The lockers are numbered from 1 to 100. When Louise gets to the lockers, they are all open. Absentmindedly, Louise closes all the even-numbered lockers—the multiples of 2—as she walks by.

A few minutes later, Jeremy comes by. He decides to touch only those lockers whose numbers are multiples of 3. If one of these lockers is open when he goes by, he closes it, and if it's closed, he opens it. (For example, Louise left locker 3 open, so Jeremy closes it. Louise closed locker 6, so Jeremy opens it, and so on.)

Then another student comes by, and this student changes the doors on all the lockers whose numbers are multiples of 4. Then another student changes the doors on lockers whose numbers are multiples of 5, and so on, until finally a student comes by who changes only locker 100.

The question is,

*Which lockers are open at the end of the process?*

You should not only determine which lockers end up open but also find an explanation for the result. Once you're done, explain which lockers would end up open if the locker numbers went up to 1000. (Assume that the last student changes only locker 1000.)

# Who's Got an Equivalent?

For each of the expressions below:

- Find an expression without parentheses that is equivalent to the given one.

- Explain why the two expressions are equivalent. (Thinking about the hot-and-cold-cube model may help.)

1. $12 - (a + 7)$

2. $26 - (12 - 3t)$

3. $41 - 2(b + 1)$

# Make It Simple

The task of removing parentheses from an expression being subtracted is a tricky one, and not easy to explain. When the expression in parentheses itself involves subtraction (as in Question 2 of *Who's Got an Equivalent?*), it's even harder.

1. Describe and explain the steps involved in simplifying these expressions.

    a. $20 - 5(x + 3)$

    b. $20 - 5(x - 3)$

2. The next expression can be simplified all the way to $4x + 7$. Show the process of simplifying it to that point.

$$6(3x + 4) - 4(x - 2) - 5(2x + 5)$$

# *Linear in a Variable*

Much of your work in *Solve It!* has involved linear equations and linear expressions. Some of the ideas for working with linear equations can be applied to equations that are not linear. An important case involves equations and expressions that include more than one variable but that are linear "in a particular variable."

For example, consider the expression $4u + tu + 9$. This expression is not linear, because it involves the product of two variables, $t$ and $u$. But if $t$ were replaced by a specific number, the expression would become linear. For instance, substituting 3 for $t$ gives $4u + 3u + 9$, which is equivalent to the linear expression $7u + 9$.

The expression $4u + tu + 9$ is called **linear in *u*,** because replacing the other variable by a number gives a linear expression. You can use the distributive property to see that this expression is equivalent to $(4 + t)u + 9$. You can think of the sum $4 + t$ as the coefficient of $u$.

1. Solve the equation $4u + tu + 9 = 6t - 4$ for $u$ in terms of $t$.

2. Solve each of the equations below for the variable indicated. In each case, the equation is linear in that variable, even if the equation as a whole is not linear.

   a. Solve for $c$ in terms of $a$.

   $$ac + 3c = 5a - (c + 7)$$

   b. Solve for $w$ in terms of $u$ and $v$.

   $$4uw + v = 3w + v^2$$

# The Shadow Equation Revisited

Do you remember this picture from *Shadows* in Year 1? The small triangle and the big triangle in this diagram are similar, so the variables $S$, $D$, $H$, and $L$ satisfy the equation

$$\frac{S}{S + D} = \frac{H}{L}$$

(Remember that $S + D$ is the length of the horizontal side of the big triangle.)

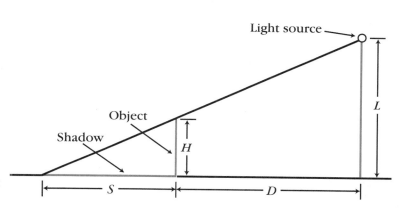

Your goal in *Shadows* was to express $S$ directly as a function of $L$, $H$, and $D$. One way to accomplish this is to add a line segment to the diagram that creates another triangle similar to the original two. This gives another equation that is easier to use in solving for $S$.

1. Your first task in this activity is to use algebra instead of a new diagram to accomplish the *Shadows* goal.

   a. Find an equation equivalent to $\frac{S}{S + D} = \frac{H}{L}$ that expresses $S$ in terms of $L$, $H$, and $D$. In other words, your equation should begin "$S = $" and have an expression involving the other three variables on the right side.
   (*Hint:* First find an equivalent equation that has no fractions in it, and then use the distributive property and factoring.)

   b. In *Lamppost Shadows*, Nelson was standing 20 feet from a 25-foot lamppost. Nelson is 6 feet tall. Use the equation you got in Question 1a to find the length of his shadow.

2. Algebra can be used to solve for other variables as well.

   a. Find an equation equivalent to $\frac{S}{S + D} = \frac{H}{L}$ that expresses $D$ in terms of $L$, $H$, and $S$.

   b. Use your equation from Question 2a to find out where Nelson should stand in order to cast a 50-foot shadow.

# *A Function—Not!*

You have occasionally used In-Out tables in contexts in which these tables did not represent functions. For example, in *The Overland Trail,* a table showed the number of people in the group in one column and the amount used of a supply item in another. This relationship wasn't a function because groups of the same size might have used different amounts of the item.

The distinction between a function and an arbitrary set of pairs is sometimes important and might be stated like this:

> The term *function* is only applicable when the *Out* is completely determined by the *In.* In other words, for an In-Out table to represent a function, there can only be one value for the *Out* for any particular choice of the *In.*

## *The Vertical-Line Test*

One way to identify which graphs are graphs of functions is to apply the **vertical-line test:**

> For a graph to represent a function, no vertical line can go through more than one of its points.

*Note:* The definition of the term *function* permits many *In* numbers to have the same *Out,* so there is no "horizontal-line test" required for functions.

1. Identify which of these graphs are graphs of functions, and explain your decisions.

   a.

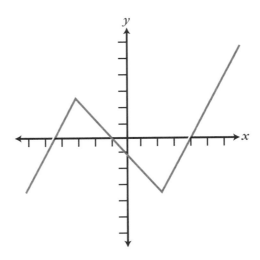

*Continued on next page*

b.

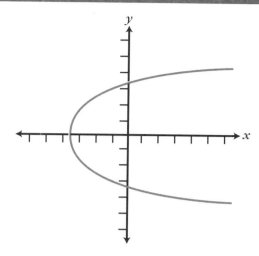

c.

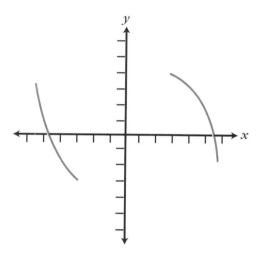

2. Explain why the vertical-line test tells you if a graph represents a function or not.

# Is There Really a Difference?

**Days 1-3**

# Data, Data, Data

This unit is about data—more specifically, about how and when to draw conclusions by comparing sets of data. In the opening days of the unit, you will generate and study some data about yourself and your classmates. You will also investigate ways to represent data in graphical form and learn about different stages in developing and evaluating hypotheses about data.

*Jahnavi Anderson, Jamie Langley, David Rodgers, and Kira Duckett use manipulatives to represent the mixed nuts in the activity "Try This Case."*

# Homework 1 Samples and Populations

## *Part I: Picking a Sample*

When you want to test a hypothesis about a population, you need to pick a sample that is likely to represent the population accurately. Consider that idea in answering these questions.

1. A music producer wants to find out what high school students think about different kinds of popular music. The producer conducts a survey among several high school Boy Scout troops to get their opinion.

   a. Do you think Boy Scouts form a good sample of high school students? If not, what might be a more representative sample? Explain your answer.

   b. Give an example of a conclusion that the producer might reach based on a survey of Boy Scouts that might not be true about high school students in general.

*Continued on next page*

2. An auto manufacturer wants to conduct on-the-street interviews to find out what adults in the United States think of the company's latest TV advertising campaign. The interviewer decides to use a group of people standing at the bus stop near her home as the sample population.

   a. Do you think the people at the bus stop form a good sample of adults in the United States? If not, what might be more representative sample? Explain your answer.

   b. Give an example of a conclusion that the interviewer might reach based on a survey of the people at the bus stop that might not be true about adults in the United States in general.

3. A member of the City Council wants to know what people in the city think about a proposed new park in the center of town. The councilmember picks names out of the city phone book at random and calls them to get a sample of opinions.

   a. Do you think names picked from the phone book form a good sample of the city's population? If not, what might be a better way to get a representative sample? Explain your answer.

   b. Give an example of a conclusion that the councilmember might reach based on a survey of people picked at random from the phone book that might not be true about the city's population in general.

## Part II: Literary Sorting

An English teacher conducted a survey, asking students about their reading preferences. She listed four book titles and asked students to identify their favorite among those choices.

She surveyed sophomores and juniors. Each sophomore puts an S next to the title of his or her favorite book on the list, and juniors put J's next to their favorites. Here are the results of the survey.

| | |
|---|---|
| *The Grapes of Wrath* | J J S J S S J J J S S S J J J J |
| *To Kill a Mockingbird* | S J S J S S J S J S S S S J S S |
| *The Bluest Eye* | S S S J S S J J S S J J J J J S J J |
| *The Great Gatsby* | J J S J S S S J J J S S S S J J J S S S |

Make a double-bar graph of this set of data.

# POW 4      *A Timely Phone Tree*

Leigh's parents were concerned about all the time she was spending on the phone. They decided to limit her calls to the period from 8:00 p.m. to 9:00 p.m. on school nights. The parents of Leigh's friends adopted the same rule.

Leigh and her friends have decided they need a more efficient way of spreading news, because long phone conversations seem out of the question. By experimenting, they have found that allowing three minutes per phone call gives them enough time to tell someone the essentials of most situations. (Three minutes includes the time needed to place the call and get the right person on the line.)

So here's their plan: The next time Leigh learns something that she thinks everyone will want to hear, she will call Mike at exactly 8:00. That call will take three minutes. Then, at 8:03, Leigh will call Diane while Mike calls Ana May. After they all finish talking, those four will each call someone else, and so on.

Your task in this POW is to figure out how many of Leigh's friends could hear the news by 9:00. In analyzing the situation, you should make certain assumptions.

- It always takes three minutes to make a connection and complete a call.

- No one calls a person who has already been called.

- The caller never gets a busy signal.

## Write-up

1. *Problem Statement*

2. *Process:* Explain how you kept track of your results.

3. *Solution*

4. *Evaluation*

5. *Self-assessment*

Interactive Mathematics Program

# Try This Case

Mr. Swenson read an ad in the newspaper for a can of mixed nuts from the Fresh Taste company. The ad claimed that "Fresh Taste Mixed Nuts contains peanuts, walnuts, almonds, and cashews in the ratio 4 : 3 : 2 : 1."

Well, Mr. Swenson was at his local supermarket in no time flat. He bought a can of Fresh Taste Mixed Nuts and went home to count. He found 300 nuts in the can— 125 peanuts, 98 walnuts, 53 almonds, and 24 cashews. Mr. Swenson wants to sue the Fresh Taste company for false advertising.

1. Imagine that you are a lawyer. What arguments would you present if you were representing Mr. Swenson? What if you were representing the Fresh Taste company?

2. Now imagine that your group is the jury on this case. Do you think there is enough evidence for Mr. Swenson to win his case? Explain why or why not. If you think he doesn't have enough evidence, explain what further evidence you would need before ruling in Mr. Swenson's favor.

# Homework 2   Who Gets A's and Measles?

People often want to know what actions might lead to certain results. One approach to finding out is to study what actions have preceded these results in the past.

Read the descriptions of two such studies and then answer the questions.

## The Situations

### Situation 1

Clarabell wanted to find out how to get an A on her next unit assessment. To do so, she had her classmates fill out a questionnaire asking about their activities before the last assessment. Specifically, she asked what they had eaten and what they had done both the night before the assessment and the morning of the assessment.

Then she found out who had gotten A's on the assessment. She found that all the "A" students had eaten dessert and listened to rap music the night before the assessment, and had drunk juice the morning of the assessment. None of the other students had done all three. (You can probably guess what Clarabell will do before the next unit assessment!)

### Situation 2

A medical researcher was trying to understand a measles epidemic in a city. She used a computer to examine the records of 500 patients of various doctors. These records showed each person's sex, blood type, cholesterol level, blood pressure, weight, height, and medical history.

The researcher saw that all the patients with measles were overweight and had high blood pressure. She concluded that overweight people with high blood pressure are more likely to contract measles.

## The Questions

1. How are the methods used in these two studies the same, and how do they differ?

2. What can be concluded from each of the studies? Is either study useful? If so, how and why?

3. How could the research methods be improved?

# Homework 3　　Quality of Investigation

## Part I: Cigarettes and Lung Cancer

The situations in *Homework 2: Who Gets A's and Measles?* illustrated the danger of jumping to conclusions. You should realize that even professional researchers often take unwarranted shortcuts. But you should not get the impression that no research study is ever legitimate or that you cannot trust the conclusions of any such study.

You may be familiar with the many studies that link cigarette smoking to lung cancer. Because of this, federal law requires that every package of cigarettes contain a warning on the connection between tobacco smoking and lung cancer. The tobacco industry has argued that this is unfair—they say that they can point to studies that do not show such a connection.

What do you think? Are such warnings unfair to the tobacco companies? How should you evaluate conflicting studies? Explain.

## Part II: Take Your Pick

Choose either Option 1 or Option 2 for Part II of this assignment.

Option 1: Do all three parts (a, b, and c).

a.  Find an article in a newspaper or magazine that reports on a particular research study or survey.

b.  Summarize the main ideas or conclusions of the study.

c.  Identify information you would like to have about the study that was not included in the article.

Option 2: Do all three parts (a, b, and c).

a.  Use your own environment—such as your school, home, or neighborhood—to gather some data.

b.  Describe a "population" for which your data could be a sample.

c.  Give an example of an *incorrect* generalization that one might make about that population from your data.

### Days 4-9

# *Coins and Dice*

Coins and dice are good tools for studying probability and data, because we have theoretical models of what to expect.

*Hazel Corbi prepares a double bar graph from the collected data.*

One of the central problems for this unit involves a coin of questionable fairness. You'll meet this coin in Situation 1 of the activity *Two Different Differences* and make your own tentative decision about whether the coin seems fair. Situation 2 of the activity involves a survey on soft drinks.

In the next few days, you'll be using examples of both coins and dice to think about random events. You'll return to the case of the questionable coin (as well as the soft drink survey) near the end of the unit.

This portion of the unit also introduces the concept of a null hypothesis—a kind of "neutral" assumption that statisticians often use in investigating data.

# Two Different Differences

Two situations are presented here, with data for each. In each case, you are asked what conclusions you can draw from the given data and how confident you are of those conclusions.

## Situation 1: A Suspicious Coin

Every time there was an extra dessert at Roberto's house, his older brother took out his special coin. He always let Roberto flip the coin and he always called out, "Heads."

It seemed to Roberto that his brother won more often than he should have with a fair coin. One day when his brother was out, Roberto found the coin and flipped it 1000 times! He got 573 heads and 427 tails.

Do you believe this was a fair coin? That is, do you think the apparent preference for heads is because the coin is not balanced, or is it just a coincidence? Your group should try to reach a consensus about this issue and then write a report that includes three items.

- A statement of Roberto's hypothesis

- A statement of your group's conclusion about the coin and an explanation of why you reached that conclusion

- A rating of how confident you are of your conclusion based on a scale of 0 to 10, with 0 meaning "no confidence" and 10 meaning "complete confidence"

*Continued on next page*

# Situation 2: To Market, to Market

A soft drink company has developed a new product. Some people in the company's marketing department think the new beverage may appeal more to men than to women. But they need to verify this so they can target their advertising properly.

They surveyed 150 people at a local supermarket to see how people liked the new product. They gave these people a sample of both their new soft drink and their current best-selling soft drink, and asked them which they preferred.

Of the 90 men questioned, 54 preferred the new soft drink and 36 liked the old one better. Of the 60 women surveyed, 33 preferred the new soft drink and 27 liked the old one better.

Based on this data, do you think the new soft drink will be more popular with men than with women? As with Situation 1, your group should try to reach a consensus and then write a report. Your report should include three items.

- A statement of the marketing department's hypothesis

- A statement of your group's conclusion about how men and women view the new beverage and an explanation of why you reached that conclusion

- A rating of how confident you are of your conclusion based on a scale of 0 to 10, with 0 meaning "no confidence" and 10 meaning "complete confidence"

# Homework 4    Changing the Difference

## Part I: Coins and Soft Drinks

Questions 1 through 5 relate to the activity *Two Different Differences*. For some of these questions, there is no single right answer, but all the questions require some thoughtfulness on your part.

1. Suppose you flip a fair coin 1000 times. What do you think is the probability of getting exactly 500 heads?

2. Roberto's data in *A Suspicious Coin* might or might not convince you that his brother's coin is unfair. Whether or not you are convinced by that data, give an example of data (for 1000 flips) that *would* convince you that the coin is unfair.

3. Suppose you find a coin and flip it 1000 times. How would you complete each of these sentences?

   a. If the experiment gives between –?– and –?– heads, then I will accept that the coin is fair.

   b. If the experiment gives between –?– and –?– heads, then I will suspect that the coin is biased in favor of heads.

*Continued on next page*

c. If the experiment gives more than –?– heads, then I will be fairly certain that the coin is biased in favor of heads.

4. In *To Market, to Market*, the survey contained 90 men and 60 women. For these questions, keep those totals but adjust the number of people within each group who prefer each drink as needed.

   a. Make up data that would make you think that the new soft drink was liked equally by men and women.

   b. Make up data that would convince you that the new soft drink was liked more by men than by women.

   c. Make up data that would convince you that the new soft drink was liked more by women than by men.

5. How does the situation in *A Suspicious Coin* differ from the situation in *To Market, to Market*? How are these situations the same?

## Part II: Stick-up Ideas

Make up a question that your class can use for a stick-up graph. This question should compare two populations in some respect and be appropriate for use with your class.

# Homework 5   Questions Without Answers

1. A record company executive is trying to convince a recording artist to go on tour to promote the artist's newest release.

   a. What do you think is the executive's hypothesis about the effect of a tour?

   b. What null hypothesis might the artist propose in order to avoid the tour?

2. A pharmaceutical company wants to advertise its new product as an anti-acne medicine. The Food and Drug Administration (FDA) is opposed to that plan.

   a. What is the company's hypothesis about its product?

   b. What null hypothesis might the FDA want to test?

3. Some of the owners in a professional basketball league want to raise the basket from ten feet to twelve feet in order to make the game more exciting.

   a. What is the hypothesis of the owners who support the change?

   b. What null hypothesis might the other owners propose?

4. A newspaper marketer is trying to persuade a business owner to place an ad in the Sunday paper.

   a. What hypothesis would the marketer suggest to the business owner?

   b. What null hypothesis might the owner believe instead?

5. Students in a Year 1 IMP class changed the weight of their pendulum's bob and got a different period for their pendulum.

   a. What hypothesis does their experiment suggest?

   b. What null hypothesis should they consider?

# Loaded Dice

Work on this task with a partner from your group. One pair from your group will make a fair die and one pair will make a loaded die. Before you begin, the group needs to decide which pair will make which kind of die.

You will be given a pattern from which to make the die. Use these steps to make your die.

1. Cut out the pattern around the outside solid line.

2. Carefully fold the pattern on all the dotted lines. The pattern will fold better if you score the lines first with a ballpoint pen.

3. Figure out how to fold the pattern into a cube with all the dots on the outside. After you have folded your die, carefully unfold it.

4. If you are "loading" a die (only one pair of students in a group should do this!), carefully tape two regular paper clips to the inside of one of the faces of the die.

5. Fold the die back together again, using cellophane tape to hold your cube together.

If time allows, roll your die many times. Are you getting an even distribution of the numbers 1 through 6?

# Homework 6                              Fair Dice

1. Roll a regular die 60 times, keeping track of the numbers that come up. Make a frequency bar graph of the data. Then repeat the process and make a separate frequency bar graph of this new data.

   a. Do the two graphs look exactly alike? If not, how are they different?

   b. Do you think your die is fair? Explain your reasoning.

2. Lucky Lou's Game Shop received a shipment of loaded and fair dice. Of course, the loaded dice are designed to look and feel just like the standard dice. Unfortunately, Lou accidentally mixed all the dice together.

   a. Suppose you were trying to figure out which were loaded and which were fair by rolling each die numerous times. How many times do you think you would have to roll each die?

   b. What kind of results would you need to see to determine whether the die was fair or loaded? Explain your reasoning.

# *Loaded or Not?*

You and your partner have been given one of the dice made yesterday. It may be a "loaded" die or it may be an "unloaded" die. You should complete Question 1 with your partner and then do the write-up in Question 2 on your own.

1. With your partner:

    a. Roll your die 60 times and make a frequency bar graph of the data.

    b. Compare the frequency bar graphs you and your partner made at home (two graphs for each of you) and the frequency bar graph you made in Question 1a.

2. On your own:

    a. Write about what you observed in making this comparison.

    b. Do you think you have a fair die or not? Explain the basis for your decision.

    c. On a scale from 0 to 10, with 0 meaning "no confidence" and 10 meaning "complete confidence," how confident are you that you correctly identified the die as fair or loaded? Explain the rating you choose.

# Homework 7      The Dunking Principle

Mr. Rose, the principal at Bayside High, agreed to participate in a dunking booth at the school fair. Here's how the booth worked.

> People who purchased tickets would push a button. The button operated a light above the booth, and the light was programmed to flash either green or red, using a randomizing mechanism. If the light turned green, Mr. Rose would fall into the water. If it turned red, he would not. He was told that the light was set to have a 50% chance of turning green.

As it turned out, Mr. Rose seemed to be getting dunked far more than 50% of the time. In fact, after 20 pushes of the button, he had been dunked 15 times! He was pretty suspicious of what he had gotten himself into.

1. What is Mr. Rose's hypothesis? What would be the null hypothesis?

2. Based on the results so far, do you think Mr. Rose is justified in being suspicious?

3. Suppose the fair continued, and Mr. Rose got dunked 46 times out of 60 pushes of the button. Would that convince you that the dunking booth was not set the way he had been told? What about 72 dunks out of 100 pushes of the button?

4. Mr. Rose's experience—15 dunks out of 20 pushes of the button—represented 75% dunks. Question 3 asked about other examples involving approximately 75% dunks.

If you were not convinced by either result described in Question 3 that the booth was not set at 50% green, how many occurrences of about 75% dunks would it take to convince you of that? If you were convinced by results like those described in Question 3, what is the *smallest* 75% dunk result that would convince you?

Explain your answer in either case.

# POW 5

# *Punch the Clock*

Imagine that you have a large number of mechanical clocks in your home. (These are traditional, analog (nondigital), 12-hour clocks.) Unfortunately, none of your clocks keeps time properly. Each clock consistently gains or loses a fixed number of minutes each hour, but different clocks may gain or lose different amounts.

At noon today, all the clocks are correct. Your task is to investigate when, if ever, the clocks will be correct again. To begin your investigation, look at these three specific examples. In each case, give as general an answer as you can.

- If a clock gains 10 minutes each hour, when will it be correct again?

- If a clock loses 3 minutes each hour, when will it be correct again?

- If a clock gains 7 minutes each hour, when will it be correct again?

Use these examples to get a feel for the situation. Then pursue the problem on your own. For example, you might look at when two different clocks will show the same time, even if neither is correct, or you might try to figure out when all possible clocks will be correct again.

Whatever you investigate, always give as general an answer as you can.

*Continued on next page*

# Write-up

Use these categories for your write-up.

1. *Subject of Exploration:* Describe the situation that you are investigating. What questions do you want to explore?

2. *Information Gathering:* Based on your notes (which should be included with your write-up), state what happened in the specific cases you examined.

3. *Conclusions, Explanations, and Conjectures:* Describe any general conclusions that you reached. Wherever possible, explain why the conclusions are true. That is, try to prove your conclusions.

   Also include statements of any conjectures you made, that is, statements that you think might be true.

4. *Open Questions:* List any questions you have that you were not able to answer. What other investigations would you do if you had more time?

5. *Self-assessment*

Adapted from *Mathematics Teacher* (Vol. 82, No. 9), © December 1989 by the National Council of Teachers of Mathematics.

# *How Different Is Really Different?*

In *A Suspicious Coin* (the problem of Roberto's brother's coin from *Two Different Differences*), you wanted to know if the coin was fair. If a fair coin were flipped 1000 times, you would expect to get about 500 heads. But Roberto got 573 heads from his brother's coin. So the question in that problem was this.

*Is this far enough from the expected number of heads to conclude that the coin is unfair?*

Similar questions are posed here for each of several different coins.

## *The Unfairest Coin*

Several students each had a coin to flip. They flipped their individual coins different numbers of times and obtained these results.

|  | Number of heads | Number of tails |
|---|---|---|
| **Alberto** | 14 | 6 |
| **Bernard** | 55 | 45 |
| **Cynthia** | 460 | 540 |

1. For each of the three students:

   a. Find the number of heads and tails that would be expected for that person's coin if the coin were fair.

   b. Compare these *expected* numbers to the numbers actually *observed*.

2. Suppose you knew that exactly one of the three coins was unfair (but didn't know which coin it was). Based on your analysis in Question 1, which coin would you suspect most? Which would you suspect least? Be sure to explain your choices and your reasoning.

# Homework 8

# Whose Is the Unfairest Die?

Alberto, Bernard, and Cynthia have some friends who prefer rolling dice to flipping coins. These friends— Xavier, Yarnelle, and Zeppa—were rolling dice recently and kept track of how many 1's they got. Each had one die to work with, and each rolled a different number of times, with these results.

|  | Number of 1's | Number of other rolls |
|---|---|---|
| **Xavier** | 1 | 29 |
| **Yarnelle** | 23 | 77 |
| **Zeppa** | 178 | 822 |

1. For each die:

   a. Find the number of 1's and the number of other rolls that would be expected if the die were fair.

   b. Compare these expected numbers to the observed numbers.

2. Suppose you knew that exactly one of the three dice was unfair (but didn't know which die it was). Based on the information from Question 1, which one would you suspect most? Which would you suspect least? Give your reasoning.

# Homework 9 Coin Flip Graph

1. Make a frequency bar graph of the class's coin flip results. Your graph should show how often heads occurred in each 100-flip set.

2. What percentage of the results had exactly 50 heads?

3. What percentage of the results had 55 or more heads?

4. About 90% of the results were between what two results?

# Days 10-18

# *A Tool for Measuring Differences*

You may recall *standard deviation* as a tool for studying measurement variation. This section of the unit is devoted to developing another tool, called the *chi-square statistic,* for evaluating experimental variation. ("Chi" is the name of a Greek letter, written as $\chi$ and pronounced like "sky" without the "s.")

*Jenna Spencer explains to Ben Holm how she calculated the chi-square statistic for each of her coin-flipping results.*

Like standard deviation, the chi-square statistic involves a complicated computation. But you'll see that learning how to calculate this statistic is just part of the task. The real challenge is learning how to use and interpret it.

# *Normal Distribution and Standard Deviation Facts*

## *What Is the Normal Distribution?*

Many phenomena have a similar distribution pattern—commonly called **bell shaped**—in which a particular result becomes gradually less likely the farther it is from the average. The **normal distribution** is a special case of a bell-shaped distribution. It has a precise mathematical definition and frequently can be used as an excellent approximation to real-world situations.

The diagram at the right shows a typical normal curve, in which the height of the curve above a given point on the horizontal axis reflects the likelihood of getting the result marked on that axis. If you draw two vertical lines on the graph, the total area under the curve between those lines gives the probability of getting a data result between the values on the horizontal axis corresponding to those lines.

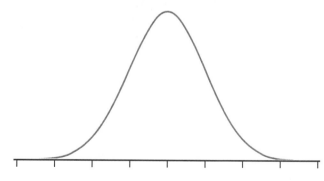

## *What Is Standard Deviation?*

The **standard deviation** of a set of data measures how "spread out" the data set is. In other words, it tells you whether the data items bunch together close to the mean or are distributed "all over the place."

The two superimposed graphs here show two normal distributions with the same mean, but the taller graph is less "spread out." Therefore, the data set represented by the taller graph has a smaller standard deviation.

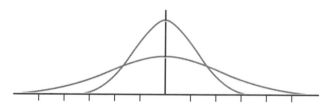

*Continued on next page*

# *Standard Deviation and the Normal Distribution*

One of the reasons the standard deviation is so important for normal distributions is that some principles about standard deviation hold true for any normal distribution. Specifically, whenever a set of data is normally distributed, these statements hold true.

- Approximately 68% of all results are within one standard deviation of the mean.

- Approximately 95% of all results are within two standard deviations of the mean.

These facts can be explained in terms of area, using the diagram "The Normal Distribution."

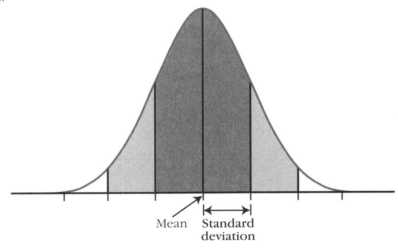

**The Normal Distribution**

In this diagram, the darker shaded area stretches from one standard deviation below the mean to one standard deviation above the mean and is approximately 68% of the total area under the curve. The light and dark shaded areas together stretch from two standard deviations below the mean to two standard deviations above the mean and constitute approximately 95% of the total area under the curve. So standard deviation provides a good rule of thumb for deciding whether something is "really different."

*Note:* In order to understand exactly where the specific numbers "68%" and "95%" come from, you would need to have a precise definition of *normal distribution*, a definition that is stated using concepts from calculus.

*Continued on next page*

# Geometric Interpretation of Standard Deviation

Geometrically, the standard deviation for a normal distribution turns out to be the horizontal distance from the mean to the place on the curve where the curve changes from being concave down to concave up. In this diagram, "Visualizing the Standard Deviation," the center section of the curve, near the mean, is concave down, while the two "tails" (that is, the portions farther from the mean) are concave up.

The two places where the curve changes its concavity, marked by the vertical lines, are exactly one standard deviation from the mean, measured horizontally.

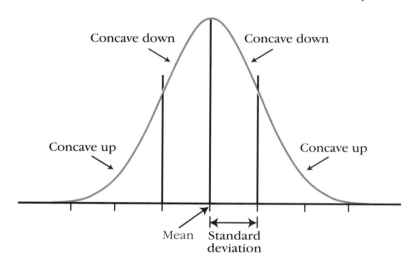

**Visualizing the Standard Deviation**

# Calculation of Standard Deviation

Here is a list of the steps for calculating standard deviation.

1. Find the mean.

2. Find the difference between each data item and the mean.

3. Square each of the differences.

4. Find the average (mean) of these squared differences.

5. Take the square root of this average.

To calculate the standard deviation, it can be very helpful to organize the set of data into a table like the one shown on the next page. The table is based on a data set of

*Continued on next page*

five items: 5, 8, 10, 14, and 18. The mean for this data set is 11. The mean of a set of data is often represented by the symbol $\bar{x}$, which is read as "*x* bar."

The computation of the mean is shown below the table to the left. Step 4 of the computation of the standard deviation, shown below the table to the right, is broken down into two substeps: (1) adding the squares of the differences and (2) dividing by the number of data items.

The symbol usually used for standard deviation is the lower case form of the Greek letter *sigma,* written as σ.

| $x$ | $x - \bar{x}$ | $(x - \bar{x})^2$ |
|---|---|---|
| 5 | -6 | 36 |
| 8 | -3 | 9 |
| 10 | -1 | 1 |
| 14 | 3 | 9 |
| 18 | 7 | 49 |

sum of the data items = 55

number of data items = 5

$\bar{x}$ (mean of the data items) = 11

sum of the squared differences = 104

mean of the squared differences = 20.8

σ (standard deviation) = $\sqrt{20.8} \approx 4.6$

Suppose you represent the mean as $\bar{x}$, use *n* for the number of data items, and represent the data items as $x_1, x_2$, and so on. Then the standard deviation can be defined by the equation

$$\sigma = \sqrt{\frac{\sum_{i=1}^{n}(x_i - \bar{x})^2}{n}}$$

# Bacterial Culture

Scientists at a research lab are trying to determine whether a new product called "Infect-Away" is better at killing bacteria than "Bact-Out," which is the current leader in the market.

The researchers have already tested Bact-Out extensively, applying it to a bacterial culture of a fixed size and measuring the number of bacteria left after a fixed time. They have seen that the number of bacteria left appears to be normally distributed. The average number of bacteria left (using Bact-Out) is 1000, and the standard deviation is 50.

1. Sketch this normal distribution, showing the mean and the approximate location of the first and second standard deviation borders.

2. How would you fill in the blanks in this sentence?

    If the bacterial test was performed over and over, one would expect about 95% of the results to show between –?– and –?– bacteria left after Bact-Out was applied.

In order to evaluate Infect-Away, the researchers use the same procedure as for Bact-Out.

3. a. What is their null hypothesis?

   b. What might their hypothesis be?

4. In testing Infect-Away, suppose they find that the culture has 825 bacteria left. What should they conclude and why?

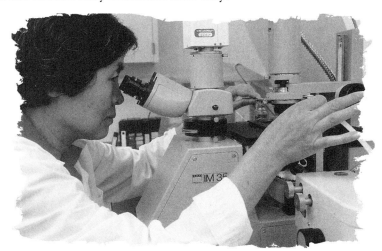

# Homework 10    Decisions with Deviation

1. The Typhoon Twister water ride has a minimum height requirement of 5 feet. The average height of children who visit this amusement park is 4 feet 10 inches, with a standard deviation of 2 inches.

   Approximately what percentage of the children who come to the park qualify to ride the Typhoon Twister? Explain. (Assume that the distribution of children's heights is approximately normal.)

2. Lob and Smash is a company that manufactures tennis balls. The company wants to get a contract to provide the Big Bucks Tennis League with tennis balls for its tournaments.

   The tennis league wants tight control over the bounce of the tennis balls it uses. The league has a standard "bounce test," and it wants balls that bounce approximately 3 feet (when dropped from a certain height), varying no more than 2 inches from this amount.

   The bounce of the Lob and Smash tennis balls is approximately normally distributed. Using the league's test, these tennis balls were shown to have a mean bounce height of 2 feet, 11 inches, with a standard deviation of 0.5 inch.

   Should the Big Bucks Tennis League use Lob and Smash tennis balls? Why or why not?

3. The time it takes Sam to get out of bed, get dressed, eat, and catch the bus varies from morning to morning. Measuring from the moment she wakes up until she reaches the bus stop, her routine takes an average of 15 minutes and has a standard deviation of 5 minutes. (Assume that the duration of her routine is normally distributed.)

   Sam likes to catch the 7:00 a.m. bus so that she can arrive early at school and visit with her friends. What time should she wake up in order to catch this bus about 84% of the time?

# Homework 11    The Spoon or the Coin?

Roberto wasn't sure whether his brother's coin was fair or not, but he certainly didn't like the results he had obtained in the past. He told his brother that he would no longer have the fate of the extra dessert determined by that coin.

Roberto's brother chuckled and gave Roberto a new option. They would toss ten spoons in the air and let them fall on the floor.

- If more than five of the spoons landed "bowl down," then Roberto would get the extra dessert.

- If more than five of the spoons landed "bowl up," then Roberto's brother would get the extra dessert.

- If the spoons were evenly split—five up and five down— then they would toss the spoons again.

(*Note:* The bowl of a spoon is the part that holds food. A spoon is "bowl up" if it would hold soup and "bowl down" if it would not.)

1. Put yourself in Roberto's place and conduct a set of experiments that will help you decide whether to use this method for deciding who gets the extra dessert.

2. What would you conclude from your experiments? How confident are you of your conclusions?

# *Measuring Weirdness*

The results from Roberto's brother's coin made Roberto suspect that it wasn't a fair coin.

Standard deviation is one tool for measuring how unusual a result is, especially in a situation involving a normal (or almost normal) distribution. But you don't always know the standard deviation for a situation, and many distributions are not normal. Therefore, other "measures of weirdness" are needed besides standard deviation.

In this assignment, you will look at ways to measure "weirdness" for coin flip results.

Recall the data of the three coins in the activity *How Different Is Really Different?*

|  | Number of heads | Number of tails |
|---|---|---|
| **Alberto** | 14 | 6 |
| **Bernard** | 55 | 45 |
| **Cynthia** | 460 | 540 |

In that activity, you compared each student's "observed numbers" to the "expected numbers" and decided which coin seemed to be the most "suspicious." In other words, you investigated which coin result would be the "weirdest" if you had a fair coin.

*Continued on next page*

Well, Alberto and Bernard have each come up with their own way to measure the "weirdness" of the coin flips.

1. Alberto's method is to find the numerical difference between the observed and the expected number of heads. Because he doesn't care whether the weirdness shows up as too many heads or as too few, he subtracts the smaller number from the larger one. For example, with Cynthia's coin, he subtracts 460 from 500 to get 40.

   a. Using Alberto's method, rank the three coins from most to least weird.

   b. Explain, using examples, why Alberto's method is inadequate for deciding which of two results is "more unusual." In other words, create two possible coin-flipping results for which the one that seems obviously "weirder" actually has a smaller numerical difference. (*Hint:* Use different sample sizes for the two experiments.)

2. Bernard's method is to use the percentage difference. That is, he finds the difference between the observed percentage of heads and the expected percentage of heads. For example, with Cynthia's coin, the percentage difference is 4%, because 46% of her flips were heads while the expected value was 50% heads.

   a. Using Bernard's method, rank the three coins from most to least weird.

   b. Explain, using examples, why Bernard's method is also inadequate for deciding which of two results is "more unusual." In this case, create two possible coin-flipping results for which the one that seems obviously "weirder" actually has a smaller percentage difference. (*Hint:* Again, use different sample sizes.)

3. Try to come up with a third method that Cynthia could use for measuring "weirdness" that you think is better than either Alberto's or Bernard's method.

# Homework 12

# Drug Dragnet: Fair or Foul?

People in many occupations, such as police officer or air traffic controller, are subject to random drug testing, on the grounds that their work affects the safety of the general public.

However, there is considerable controversy about such testing. One objection concerns the potential unfair consequences of the fact that the tests are not perfect. For instance, if the test incorrectly shows someone to be a drug user (a "false positive"), that person could lose his or her job. This assignment explores some of the mathematical issues in drug testing.

In this assignment, you are to assume that a certain test for drug use is 98% accurate. By this, we mean that 98% of the people who use the given drug within some specified time period will test positive and 98% of the people who did not use the drug in that time period will test negative. Also assume that only 5% of the people on the job (1 in every 20) engage in drug use.

1. If someone tests positive, how likely is it that the person has actually engaged in drug use within the given time period? (*Hint:* Consider a large population, such as 10,000 people. Figure out how many use drugs and how many users and nonusers test positive.)

2. Do you think a test such as this should be used? Explain.

# How Does $\chi^2$ Work?

You have seen that the $\chi^2$ ("chi-square") statistic is computed using the expression

$$\frac{(\text{observed} - \text{expected})^2}{\text{expected}}$$

Specifically, this expression is computed for each observed number. The $\chi^2$ statistic is defined to be the sum of these results. For example, for coin-flipping problems, there are two observed numbers—the number of heads and the number of tails.

Now that you know how to do the calculations, you need to know how it reflects the actual situation. What does this magic number tell you?

Your group is to investigate what different $\chi^2$ numbers mean by making up coin flip results and calculating the $\chi^2$ statistic for each.

1. Pick a sample size and make up some different coin flip results that might occur for that sample size. Calculate the $\chi^2$ statistic for each of these results. What happens to the value of the $\chi^2$ statistic as the results get "weirder"?

*Continued on next page*

You saw in *Measuring Weirdness* that Alberto's and Bernard's methods—numerical difference and percentage difference—didn't necessarily tell you which of two results was "weirder" if the sample sizes were different. The next two questions focus on whether the $\chi^2$ statistic is sensitive to changes in sample size.

2. Consider these two cases.

    - You flip a coin 1000 times and get 510 heads.

    - You flip a coin 30 times and get 25 heads.

    In both cases, you have 10 heads more than expected. So Alberto's method would rate them as "equally weird."

    a. Find the $\chi^2$ statistic for each of these two situations.

    b. Do the $\chi^2$ statistics reflect your intuition about which is "weirder"?

3. Consider these two cases.

    - You flip a coin 12 times and get 3 heads.

    - You flip a coin 60 times and get 15 heads.

    In both cases, you have 25% fewer heads than expected. So Bernard's method would rate them as "equally weird."

    a. Find the $\chi^2$ statistic for each of these two situations.

    b. Do the $\chi^2$ statistics reflect your intuition about which is "weirder"?

4. What numerical value for the $\chi^2$ statistic would lead you to believe that a coin flip result was really unusual? Explain.

# Homework 13                    The Same χ²

Getting 40 heads and 10 tails out of 50 flips of a fair coin would be pretty weird, right? Well, it turns out that the χ² statistic for that result is 18, because the expected number for both heads and tails is 25, and

$$\chi^2 = \frac{(40-25)^2}{25} + \frac{(10-25)^2}{25} = 18$$

This homework is about other coin flip results that would give you a χ² statistic of 18.

1. Suppose you flipped a fair coin 100 times.

   a. Decide, *based on your intuition*, how many heads and how many tails out of 100 flips would seem just as weird as getting 40 heads and 10 tails out of 50 flips.

   b. Calculate the χ² statistic for that 100-flip result.

   c. By trial and error, find the number of heads and tails out of 100 flips that actually comes closest to having a χ² statistic of 18.

2. Repeat all three parts of Question 1 for 1000 flips.

3. What generalizations can you come up with based on your answers to Questions 1 and 2?

# Measuring Weirdness with $\chi^2$

In *Measuring Weirdness*, you looked at Alberto's and Bernard's methods for deciding which of the coins from *How Different Is Really Different?* was the "weirdest." In this activity, you'll look first at what the $\chi^2$ statistic has to say about those coins and then at the case of *A Suspicious Coin* (Roberto's brother's coin).

For your convenience, here again are the results of the three coins.

|          | Number of heads | Number of tails |
|----------|-----------------|-----------------|
| **Alberto**  | 14              | 6               |
| **Bernard**  | 55              | 45              |
| **Cynthia**  | 460             | 540             |

1. For each coin:

   a. Find the number of heads and tails that would be expected if the coin were fair.

   b. Calculate the $\chi^2$ statistic. (*Reminder:* For coin-flipping problems, there are two observed numbers—the number of heads and the number of tails.)

2. Suppose you knew that one of the three coins was unfair. Based on your analysis in Question 1, which coin would you suspect most? Which would you suspect least? Explain your answers using $\chi^2$ statistics.

3. Calculate the $\chi^2$ statistic for Roberto's brother's coin (573 heads out of 1000 flips). Do you think that coin is unfair? Explain.

# Homework 14 χ² for Dice

In *Measuring Weirdness with* $\chi^2$, you applied the $\chi^2$ statistic to some earlier coin problems. In this assignment, you will apply the $\chi^2$ statistic to the dice data from *Homework 8: Whose Is the Unfairest Die?* and see how well $\chi^2$ fits your intuition about "weirdness" of dice results.

To refresh your memory, three friends—Xavier, Yarnelle, and Zeppa—were rolling dice and keeping track of how many 1's they got. They each had a different die, and they rolled their dice different numbers of times, with these results.

|  | Number of 1's | Number of other rolls |
|---|---|---|
| **Xavier** | 1 | 29 |
| **Yarnelle** | 23 | 77 |
| **Zeppa** | 178 | 822 |

1. For each die:

   a. Find the number of 1's and the number of other rolls that would be expected if the die were fair.

   b. Calculate the $\chi^2$ statistic.

2. Suppose you knew that one of the three friends had an unfair die. Based on the information from Question 1, which one would you suspect most? Which would you suspect least? Give your reasoning.

3. How do the $\chi^2$ results compare to your intuition?

# Does Age Matter?

Clementina is a student who waits on tables at an ice cream store after school. She earns the minimum wage, so her income is mostly from tips.

Most of her customers are high school students. She thinks that adults are more likely to be good tippers than high school students because adults have more money to spend and have a better idea of how much to tip.

Clementina's mother has waited on tables for many years at various restaurants in which all her customers were adults, and she kept track of how many tipped well. She says that in her experience, 70% of adult customers tip well.

Clementina has just been offered a job waiting on tables in a coffee shop with nearly all adult customers. She is trying to decide if she should take the new job. It's been fun working around high school students, but she thinks she might make more money if she takes the new job.

She decides to check out her theory about tipping. She keeps careful track of her tips for a day at the ice cream store and finds that out of 52 high school customers, she gets 30 good tips and 22 poor ones.

1. What is Clementina's hypothesis? What is the null hypothesis?

2. What would have been the expected number of good tips for Clementina's survey if high school customers had the same tipping habits as the adult customers in Clementina's mother's experience?

3. Calculate the $\chi^2$ statistic for Clementina's data.

4. Do you think that Clementina would earn more in tips if she changed jobs? Why?

5. If you were Clementina, would you switch? Explain your reasoning.

# Homework 15

# Different Flips

The purpose of this assignment is to begin gathering data on the probability of getting certain values of the $\chi^2$ statistic if the null hypothesis is actually true. Knowing such probabilities will allow you to use the $\chi^2$ statistic to decide whether to accept or reject a null hypothesis.

*Important:* Please do not "fake" this data set. Because you will be using this data set to find the approximate probability of getting certain $\chi^2$ statistics, it is important to have authentic information on the variety of results that actually occur.

1. Flip a coin 50 times and record the number of heads and tails you get. Write down how many heads and tails you would expect if the coin were fair. Find the $\chi^2$ statistic for your data. (Round your answer to the nearest hundredth.)

2. Do a similar experiment, but this time with 40 flips, and come up with a second $\chi^2$ statistic. (Round your answer to the nearest hundredth.)

3. Do another experiment, but this time with 60 flips, and come up with a third $\chi^2$ statistic. (Round your answer to the nearest hundredth.)

4. Are your $\chi^2$ statistics the same? Why or why not?

# Graphing the Difference

1. Make a frequency bar graph of the whole class's data from *Homework 15: Different Flips.*

2. What percentage of the class's data had a $\chi^2$ statistic greater than 1?

3. Based on your graph, what would you estimate the probability of getting a $\chi^2$ statistic greater than 3 to be?

4. What would you estimate the probability of getting a $\chi^2$ statistic less than 4 to be?

# Homework 16     Assigning Probabilities

1. In *Measuring Weirdness with* $\chi^2$, you should have found that the $\chi^2$ statistic for Bernard's coin was 1 (exactly). In *Graphing the Difference*, you made a frequency bar graph of $\chi^2$ statistics from your class's data.

   a. How many data items were used altogether to make your frequency bar graph?

   b. How many of these data items were at least as large as Bernard's $\chi^2$ statistic?

   c. Use your answers to Questions 1a and 1b to estimate the probability of getting a $\chi^2$ statistic at least as large as Bernard's.

2. Write general instructions explaining to someone how to use your $\chi^2$ frequency bar graph to estimate probabilities of coin flip results.

3. Compare the $\chi^2$ statistic to standard deviation. How are they the same? How are they different?

# *Random but Fair*

This activity should be done with a partner.

In this activity, you will be using the random number generator of your graphing calculator to simulate the null hypothesis from *Does Age Matter?* As you may recall from that assignment, Clementina's mother, an experienced server, said that 70% of adults are good tippers. The null hypothesis was that high school students fit the same model. So what you want is something like a coin that comes up heads 70% of the time.

A random number generator can act just like that coin, in this fashion.

- If the first digit of the random number (that is, the number in the tenths place) is 0, 1, or 2, it will mean that the person was not a good tipper.

- If the first digit is 3, 4, 5, 6, 7, 8, or 9, then the person was a good tipper.

In the long run, this method should give good tips 70% of the time.

Now you're ready for the activity.

1. In the first experiment, you will generate 40 random numbers. (This is like having a sample of 40 people.) You can have one partner handle the graphing calculator and report the random numbers, while the other partner records the result.

   With each random number, the person with the calculator will say "bad" if the digit in the tenths place is 0, 1, or 2, and "good" if the tenths digit is 3, 4, 5, 6, 7, 8, or 9. The partner should keep a tally of how many good and bad tippers there are, and announce when 40 results have been tallied.

2. Calculate the $\chi^2$ statistic for your result (using the 70% figure to compute your expected numbers).

3. Repeat Questions 1 and 2 for a sample of 50 people and then for a sample of 60 people.

4. Switch roles of reporter and recorder, and repeat Questions 1 through 3. When you are finished, you and your partner should have calculated a total of six $\chi^2$ statistics.

# A χ² Probability Chart

| Value of the χ² statistic | Probability of getting a χ² statistic that large or larger when the null hypothesis is true |
|:---:|:---:|
| 0.0 | 1.0000 |
| 0.2 | .6547 |
| 0.4 | .5271 |
| 0.6 | .4386 |
| 0.8 | .3711 |
| 1.0 | .3173 |
| 1.2 | .2733 |
| 1.4 | .2367 |
| 1.6 | .2059 |
| 1.8 | .1797 |
| 2.0 | .1573 |
| 2.2 | .1380 |
| 2.4 | .1213 |
| 2.6 | .1069 |
| 2.8 | .0943 |
| 3.0 | .0832 |
| 3.2 | .0736 |
| 3.4 | .0652 |
| 3.6 | .0578 |
| 3.8 | .0513 |
| 4.0 | .0455 |
| 4.2 | .0404 |
| 4.4 | .0359 |
| 4.6 | .0320 |
| 4.8 | .0285 |
| 5.0 | .0254 |
| 5.2 | .0226 |
| 5.4 | .0201 |
| 5.6 | .0180 |
| 5.8 | .0160 |
| 6.0 | .0143 |
| 6.2 | .0128 |
| 6.4 | .0114 |
| 6.6 | .0102 |
| 6.8 | .0091 |
| 7.0 | .0082 |
| 7.2 | .0073 |
| 7.4 | .0065 |
| 7.6 | .0058 |
| 7.8 | .0052 |
| 8.0 | less than .005 |

# Homework 17     A Collection of Coins

1. Al and Betty have two coins. One coin is fair and the other is not, and they are trying to figure out which is which. Betty flips the first coin for a while and gets 40% heads.

   Now it's Al's turn. He flips the other coin a few times, but then stops, thinks, and does some arithmetic. Even though Al has flipped heads only 25% of the time, he now claims that he is quite sure the coin he is flipping is the fair coin—and he's right! How can this be?

2. Find the $\chi^2$ statistic for each of the three samples of coin flips given in items a, b, and c. Then use the information in *A $\chi^2$ Probability Chart* to find the probability that you would get a result that far or farther from what you would expect if the coin were fair.

   a. 25 heads and 35 tails

   b. 220 heads and 240 tails

   c. 995 heads and 1025 tails

3. Make up a sample that has the same difference between the number of heads and the number of tails as in Question 2a (namely, a difference of 10) but that has a smaller $\chi^2$ statistic than that for Question 2a. (You do not have to actually calculate the $\chi^2$ statistic if you can explain clearly how you know that it is smaller.)

4. Make up a sample that has the same number of total flips as in Question 2c (namely, 2020 total flips) but that has a larger $\chi^2$ statistic than that for Question 2c. (You do not have to actually calculate the $\chi^2$ statistic if you can explain clearly how you know that it is larger.)

# POW 6

# *Is There Really a Difference?*

In this POW, you will make a hypothesis about two populations that you think are different in some respect and collect data from samples of each of the two populations. You will then examine the sample data and use your analysis to evaluate your hypothesis.

You should work on this POW with a partner. Together, you and your partner will be creating a poster of your results and making a presentation to the class. Each of you will also hand in a write-up of the POW.

## *Stages of the POW*

Here are the different stages of your work on this POW.

1. Tell your teacher who your partner is.

2. You and your partner hand in

   • a precise statement of your hypothesis

   • a statement of your null hypothesis

   • a method for collecting sample data from each of your populations
     (If your method of collecting data involves a questionnaire, you should
     include a copy.)

*Continued on next page*

3. You and your partner collect your data. You should take care that your sample is not biased, that participants did not influence one another, and that you got accurate information.

4. You and your partner analyze the results and prepare both a poster and a presentation.

5. You do your own write-up of the POW. (The parts of your write-up are described below.)

6. You and your partner make your presentation using your poster.

## Presentation Poster

Your presentation should include a poster that summarizes your analysis. This poster should contain these items.

- The question you investigated

- The data you collected

- Your chart of observed and expected values

- The calculation of the $\chi^2$ statistic

- The probability associated with your $\chi^2$ statistic

- Your conclusions

## Write-up

1. *Statement of Hypothesis:*

   - Explain why you chose your hypothesis. That is, explain why you thought it was true before you collected data on it.

   - State the null hypothesis.

2. *Collection of Data:* Describe how you went about collecting data. In particular, answer these three questions.

   - How did you try to guarantee that your sample would be representative of the entire population concerned in your hypothesis? That is, how did you try to avoid "bias"?

   - What did you do to keep participants from influencing one another? That is, how did you keep participants "independent"?

   - Do you think people were willing to give you accurate information, or did they tend to lie because they were embarrassed or wanted to act cool?

*Continued on next page*

3. *Analysis:* On the basis of your sample data, do you think your hypothesis is true, false, or still not proven? Explain your reasoning.

The important part of this section of your report is your reasoning. Your reasoning should be based on an analysis of the data you collected. In your analysis, you should include a double-bar graph and a graph showing where your $\chi^2$ statistic fits on the $\chi^2$ distribution curve.

4. *Evaluation:* What did you learn from doing this POW? Include both what you learned about the process of analyzing data and what you learned concerning the topic you studied.

# Homework 18

# Late in the Day

The manager of a manufacturing company is concerned about the number of on-the-job accidents and the times at which they happen. She suspects that more accidents happen at the end of a shift because workers get tired.

If this is so, she will consider changing work schedules in some way to improve safety, but schedule changes are costly and are not popular with most workers.

At this company, each employee works an eight-hour shift. The manager studied recent records and found that of the company's last 168 accidents, 57 had occurred in the last two hours of workers' shifts.

1. The manager's hypothesis is that there is a higher accident rate late in a shift. What is the null hypothesis?

2. How many accidents would you expect to have happen in the last two hours of shifts if the accidents were equally distributed throughout shifts?

3. Imagine that you are a statistician hired by the manager to advise her. Do a $\chi^2$ test with her data, and on the basis of your findings, advise her whether to change schedules. Be sure to explain your reasoning carefully to her, because she has never studied statistics.

4. What else should be considered in analyzing this situation?

# Days 19–23

## Comparing Populations

Coins and dice may be fun to play with, but many of us find it more interesting to study people. In this section of the unit, you'll see how to use the chi-square statistic to compare two populations, and you'll return to the two central problems from *Two Different Differences*.

You'll also work on *POW 6: Is There Really a Difference?* For this major project, you will gather data and apply many of the ideas of the unit.

*Once they agree on the null hypothesis, Karinn Pearson, Malikha Gardner, John Dowd, and Iasha Day proceed with work on "What Would You Expect?"*

# What Would You Expect?

Consider this question:

*Who is more likely to be left-handed—a man or a woman?*

You probably don't know the answer. In this activity, you will look at what you would expect from a survey *if there were no difference* between men and women in this respect.

In other words, your null hypothesis is that men and women are the same with respect to "handedness." You should imagine that you are taking a survey to help you decide whether to reject this null hypothesis.

Throughout this activity, you should complete the tables as if the null hypothesis were true.

1. Assume that you have a group with 100 women and 100 men. In this group of 200, there are 150 right-handed people and 50 left-handed people.

   Make a chart like the one shown here, and fill in the empty boxes with the number of people you'd expect to find in each of the four groups if the null hypothesis were true, that is, if gender made no difference to "handedness."

|              | Women | Men | Total |
|--------------|-------|-----|-------|
| **Right-handed** |       |     | 150   |
| **Left-handed**  |       |     | 50    |
| **Total**        | 100   | 100 | 200   |

*Continued on next page*

2. Assume that you have a different group of people, this one containing 120 women and 80 men. In this group of 200, there are 150 right-handed people and 50 left-handed people.

   Again, make a table like the one shown here, and fill in the number of people you'd expect to find in each of the four groups (still assuming that the null hypothesis is true).

| | Women | Men | Total |
|---|---|---|---|
| **Right-handed** | | | 150 |
| **Left-handed** | | | 50 |
| **Total** | 120 | 80 | 200 |

3. Assume that you have a third group of people, this time consisting of 25 women and 75 men. In this group of 100 people, 65 are right-handed and 35 are left-handed.

   Again, make a table like the one shown here, and fill in the number of people you'd expect to find in each of the four groups (still assuming that the null hypothesis is true).

| | Women | Men | Total |
|---|---|---|---|
| **Right-handed** | | | 65 |
| **Left-handed** | | | 35 |
| **Total** | 25 | 75 | 100 |

4. Now invent your own group of people. That is, choose how many men, how many women, how many right-handed people, and how many left-handed people there are. (Remember that because you are considering one overall group of people, the total number of men and women must be the same as the total number of left-handed and right-handed people.)

   Then make a table like the ones in the other problems and fill in the number of people you'd expect to find in each of the four groups.

   As in Questions 1 through 3, assume that the null hypothesis is true, that is, that gender makes no difference to "handedness."

# Homework 19

# Who's Absent?

Suppose that in a certain high school, an administrator forms the hypothesis that there is a difference between 10th graders and 11th graders with respect to their absences from school.

For each problem, make a table like the one given and fill it out based on the assumption that the null hypothesis is true. In other words, throughout this assignment, you are to assume that there is no difference between 10th graders and 11th graders with respect to the likelihood of being absent.

*Continued on next page*

1. Assume that there are 200 tenth graders and 200 eleventh graders in the school. Out of those 400 students, 40 are absent. Show how many students you would expect to find in each of the four groups.

|  | Absent | Not absent | Total |
|---|---|---|---|
| **10th graders** |  |  | 200 |
| **11th graders** |  |  | 200 |
| **Total** | 40 | 360 | 400 |

2. Suppose instead that there are 200 tenth graders and 300 eleventh graders in the school. This time, 60 students are absent. Show how many students you would expect to find in each of the four groups.

|  | Absent | Not absent | Total |
|---|---|---|---|
| **10th graders** |  |  | 200 |
| **11th graders** |  |  | 300 |
| **Total** | 60 | 440 | 500 |

3. Choose your own values for the number of 10th graders and number of 11th graders. Also, decide on the fraction of the total group that is absent.

   Make a table like the ones in the previous problems. Fill in the number of 10th graders and 11th graders that you would expect to be absent and the number of 10th graders and 11th graders that you would expect not to be absent, based on the assumption that grade level makes no difference in the absence rate.

# Big and Strong

The doctors in the problems in this activity have made hypotheses about the effects of the care they give their patients.

*Note:* The answers to these questions will be needed in *Homework 20: Delivering Results.*

1. Dr. Eileen Bertram is an obstetrician. She thinks that she gives better prenatal care to her patients than most doctors do and that, as a result, the babies she delivers are less likely to be underweight at birth.

   Last year she delivered 75 babies. Other obstetricians in her building delivered 280 babies. Out of the total of 355 babies, 43 were underweight.

   a. State an appropriate null hypothesis for Dr. Bertram's theory.

   b. Copy the table shown here and use the numbers given in the problem to fill in the totals (including the number of babies altogether).

|  | Underweight babies | Babies not underweight | Total |
|---|---|---|---|
| Dr. Bertram |  |  |  |
| Other doctors |  |  |  |
| Total |  |  |  |

   c. Fill in the four other boxes of the table by finding the expected numbers for this problem. In other words, figure out what the table would look like if the results were to fit the null hypothesis exactly.

*Continued on next page*

2. Dr. Oliver Pine is a pediatrician. He has kept track of the children he and his colleagues have cared for, and thinks his own patients are more likely to turn out to be good athletes than his colleagues' patients.

Altogether, he has kept track of 121 of his own patients and 348 of his colleagues' patients. Out of the total of 469 children, 217 became good athletes.

a. State an appropriate null hypothesis for Dr. Pine's theory.

b. Copy the table shown here and use the numbers given in the problem to fill in the totals (including the number of children altogether).

|               | Good athletes | Not good athletes | Total |
|---------------|---------------|-------------------|-------|
| **Dr. Pine**  |               |                   |       |
| **Other doctors** |           |                   |       |
| **Total**     |               |                   |       |

c. Fill in the four other boxes of the table by finding the expected numbers for this problem.

# Homework 20

# Delivering Results

In *Big and Strong,* you considered two situations, each comparing the patients of one doctor with the patients of his or her colleagues.

In each case, you wrote a null hypothesis and then completed tables to show what would be expected if the null hypothesis were true.

In this assignment, you will make up results to complete the tables in a way that would support a claim *different* from the null hypothesis. The totals for each table are already entered for you.

1.  This problem is based on Question 1 of *Big and Strong.* Recall Dr. Bertram's belief that the babies she delivered were less likely to be underweight at birth than the babies delivered by other doctors.

*Continued on next page*

Make up numbers for the missing entries in the table that would support her theory. Be sure that your numbers fit the totals.

|  | Underweight babies | Babies not underweight | Total |
|---|---|---|---|
| Dr. Bertram |  |  | 75 |
| Other doctors |  |  | 280 |
| Total | 43 | 312 | 355 |

2. This problem is based on Question 2 of *Big and Strong*. Recall Dr. Pine's belief that the children he cared for were more likely to turn into good athletes than the children cared for by other doctors.

Make up numbers for the missing entries in the table that would support his claim. Be sure that your numbers fit the totals.

|  | Good athletes | Not good athletes | Total |
|---|---|---|---|
| Dr. Pine |  |  | 121 |
| Other doctors |  |  | 348 |
| Total | 217 | 252 | 469 |

3. You've seen how the $\chi^2$ statistic is defined in situations in which a population is compared to a theoretical model. In the situations from *Big and Strong*, the comparison is between two actual populations. For example, in Question 1, you are comparing Dr. Bertram's patients with the patients of other doctors.

How do you think the $\chi^2$ statistic might be calculated in a situation like this? For example, how might you calculate the $\chi^2$ statistic for the data you made up in Question 1 of this assignment?

# Paper or Plastic?

Earlier in the unit, you used the $\chi^2$ statistic to find out if a given population fit a certain theoretical model.

In recent problems, you have been comparing two populations with each other. The central question has been whether the two populations *really* are different. Now you're ready to use the $\chi^2$ statistic to help answer such questions.

## The Situation

A checker at a supermarket thinks people who buy frozen dinners are more likely to prefer plastic bags than people who don't buy frozen dinners.

To check his theory (and to make his job more interesting), he kept track of people going through his checkout stand. Ultimately, he had a record of 2000 people: 500 who had bought at least one frozen dinner and 1500 who had not bought any frozen dinners.

Of those who bought frozen dinners, 260 requested plastic and 240 requested paper. Of those who did not buy frozen dinners, 740 requested plastic and 760 requested paper.

Do you think this set of data supports the checker's hypothesis that there really is a difference in the two groups of people with respect to their bag preference?

## The Report

Prepare a report about *The Situation*. Follow these steps.

- State what two populations are being compared.

- State your hypothesis and the null hypothesis.

- Calculate the expected numbers. In other words, decide what numbers you'd expect if the null hypothesis were true.

- Calculate the $\chi^2$ statistic.

- Use the $\chi^2$ probability chart to find the probability of getting a $\chi^2$ statistic that large or larger.

- Based on the probability, decide whether to reject the null hypothesis. That is, decide whether you believe that the two populations are really different. Explain your reasoning.

# Homework 21          Is It Really Worth It?

Feline leukemia is a deadly disease among cats, but it may be possible to protect them with a vaccine. A veterinarian is considering recommending a certain vaccine to his clients for their cats, based on what he has read about its use in another region.

Sometime after vaccination of cats had begun in that region, there was an outbreak of the disease. A report looked at data on 400 cats that were in the area of the outbreak.

It turned out that 100 of the cats had been vaccinated and 300 had not. The number in each group that actually got sick is shown in the table below.

|  | Got sick | Did not get sick | Total |
|---|---|---|---|
| **Vaccinated** | 10 | 90 | 100 |
| **Not vaccinated** | 50 | 250 | 300 |
| **Total** | 60 | 340 | 400 |

The veterinarian wants to know whether he should recommend to his clients with cats that they bring their pets in to get this vaccine.

It will be expensive for him to notify all of his clients and difficult to schedule the visits to give shots. Also, it will cost each of his clients $40, which many of them would rather not have to spend. And many cat owners would prefer not to subject their pets to the trauma of being vaccinated.

Yet this disease is the most common cause of death among cats in his practice. What do you think the veterinarian should do and why?

Be sure to state a null hypothesis and carry out a $\chi^2$ test. (Remember to use the row and column totals, not the observed numbers in the boxes, when you are finding the expected numbers.)

Also, explain why you think the $\chi^2$ statistic is or is not useful in this case.

# "Two Different Differences" Revisited

In *Two Different Differences*, you looked at two situations—one involving a possibly biased coin and one comparing men's and women's beverage preferences. At that time, you were only able to evaluate the situations intuitively. Now, you will return to those situations and use the $\chi^2$ statistic to make more informed judgments about them.

You should write a report for each of the situations. Your report should include these six items.

- A statement of the populations involved

- A hypothesis

- A null hypothesis

- An explanation of how you found the $\chi^2$ statistic

- A statement of your conclusion and how you used the $\chi^2$ statistic to reach that conclusion

- A rating of how confident you are of your conclusion based on a scale of 0 to 10, with 0 meaning "no confidence" and 10 meaning "complete confidence"

The two situations are summarized here.

## Situation 1: A Suspicious Coin

Roberto's brother had a special coin that he used whenever there was an extra dessert. Roberto's brother would always call "Heads."

Roberto suspected that the coin came up heads more often than it should, so one day when his brother was out, Roberto found the coin and flipped it 1000 times. He got 573 heads and 427 tails.

The question to examine is

*Is Roberto's brother's coin biased in favor of heads?*

*Continued on next page*

## *Situation 2: To Market, to Market*

The marketing department of a soft drink company suspected that its new beverage might appeal more to men than to women.

To test this hypothesis, the department surveyed 150 people at a local supermarket to see how they liked it in comparison with the company's old product.

Of the 90 men questioned, 54 preferred the new soft drink, and 36 liked the old one better. Of the 60 women surveyed, 33 preferred the new soft drink, and 27 liked the old one better.

The question to examine is

> *Will the new soft drink be more successful among men than among women?*

# Homework 22

# Reaction Time

Buck Adams had always heard that people should not drink alcoholic beverages and then drive a car. However, he wasn't sure whether everyone knew how significantly alcohol impaired one's driving ability.

As a service to his community, Buck conducted an experiment, using a driving simulator to test people's reflexes. He had some participants use the simulator while sober and others do so while intoxicated beyond the legal limit.

Buck recorded the number of participants in each category who "crashed" and the number who did not.

Here is a table of his results.

|  | Crashed | Didn't crash |
|---|---|---|
| **Sober** | 52 | 125 |
| **Drunk** | 66 | 88 |

1. What do you think Buck's hypothesis and null hypothesis were for this experiment?

2. Use the $\chi^2$ statistic to find the probability that his samples of sober drivers and drunk drivers would have this large a difference in driving performance if alcohol did not affect driving ability.

3. Why do you think some people still drive while intoxicated despite the abundance of warnings against it and penalties for doing so?

# *Bad Research*

Read the following "journal entry" from a "researcher." Then list all the mistakes you can find in the researcher's work and explanation of the work.

## *Personal Journal Entry 710:*

I was thumbing through my associate's files today and came across some files of students he taught at the Junior Academy. According to the records, 5 of the 50 boys in the class failed. Amazingly, there were no records of any girls failing!

Assuming there were an equal number of boys and girls, I proceeded to calculate the ex-square statistic to see if there was really a difference in their failure rates.

I first made this table.

|          | **Boys** | **Girls** | **Total** |
|----------|----------|-----------|-----------|
| **Passed** | Expected: 47.5<br>Observed: 45 | Expected: 47.5<br>Observed: 50 | 95 |
| **Failed** | Expected: 2.5<br>Observed: 5 | Expected: 2.5<br>Observed: 0 | 5 |
| **Total** | 50 | 50 | 100 |

Then I proceeded to do my calculations:

$$\frac{(45 - 47.5)^2}{45} + \frac{(50 - 47.5)^2}{50} + \frac{(2.5 - 5)^2}{5} + \frac{(2.5 - 0)^2}{0} = 0.14 + 0.13 + 1.25 + 0$$
$$= 1.52$$

When I looked this up on the probability table, I found that this ex-square statistic happened about 22% of the time.

So I conclude that 22% of boys fail more than girls. More importantly, it proves that women are smarter than men.

# Homework 23

# On Tour with $\chi^2$

The situations in this assignment come from *Homework 5: Questions Without Answers*. The numerical information in them has been made up.

1. A record company executive and a recording artist are arguing about whether the artist should go on a tour to promote the artist's newest release.

   The issue is whether there is a difference in record sales when an artist goes on tour. To address this issue, they collected some data and found that out of 50 acts that toured, 20 made a profit on their release's sales. On the other hand, of the 120 acts that did not tour, 30 made a profit.

   a. What is the hypothesis and null hypothesis in this situation?

   b. Set up a table, calculate the $\chi^2$ statistic, and find the probability associated with the $\chi^2$ statistic you get.

   c. Do you think these results are convincing enough to justify having the artist tour?

   d. What are some problems with this study, and how would you change it?

*Continued on next page*

2. The owners in a professional basketball league are arguing about whether to raise the basket from 10 feet to 12 feet in order to make the game more exciting. What the owners really want to know is whether there is a difference in attendance when the basket is raised to 12 feet.

The owners had the basket raised to 12 feet for some exhibition games, but left the basket at 10 feet for some other exhibition games. They then compared the attendance at each game with that of the previous year.

Of the 40 games with a 12-foot basket, 35 showed an increase in attendance. Of the 100 games with a 10-foot basket, 70 showed an increase in attendance.

a. What is the hypothesis and null hypothesis in this situation?

b. Set up a table, calculate the $\chi^2$ statistic, and find the probability associated with the $\chi^2$ statistic you get.

c. Do you think these results are convincing enough to justify changing the height of the basket at regular games?

d. What are some problems with this study, and how would you change it?

**Days 24–26**

# POW Studies

The final days of the unit are devoted to *POW 6: Is There Really a Difference?* You will complete your report and listen to one another's presentations.

In the last two homework assignments, you will compile your unit portfolio.

*Ceanna Shira begins her POW presentation by explaining the hypothesis she and her partner decided to investigate.*

# Homework 24

# Wrapping It Up

Your homework tonight is to complete your POW report. Meet with your partner so that you will be prepared if you have to make your presentation tomorrow.

# Homework 25

# Beginning Portfolio Selection

You've had an opportunity to use the $\chi^2$ statistic in a number of situations. Now look back over your homework and class activities for this unit.

1. Choose one problem for which you felt using the $\chi^2$ statistic was helpful in making a decision.

2. Choose another problem for which you felt it was not useful (or at least less useful).

3. Describe the kind of situation in which you feel the $\chi^2$ statistic is most useful in decision-making.

# Homework 26                    "Is There Really a Difference?" Portfolio

Now that *Is There Really a Difference?* is completed, it is time to put together your portfolio for the unit. This assignment has three parts.

- Writing a cover letter summarizing the unit

- Choosing papers to include from your work in this unit

- Discussing the role of *POW 6: Is There Really a Difference?* in the unit

## Cover Letter for "Is There Really a Difference?"

Look back over *Is There Really a Difference?* and describe the main mathematical ideas of the unit. This description should give an overview of how the key ideas were developed. In compiling your portfolio, you will be selecting some activities that you think were important in developing the key ideas of this unit. Your cover letter should include an explanation of why you selected particular items.

*Continued on next page*

# Papers from "Is There Really a Difference?"

Your portfolio for *Is There Really a Difference?* should contain these items.

- *"Two Different Differences" Revisited*

- *Homework 25: Beginning Portfolio Selection*

- *POW 6: Is There Really a Difference?*

- Another Problem of the Week
  Select one of the first two POWs you completed during this unit (*A Timely Phone Tree or Punch the Clock*).

- Other high-quality work
  Select one or two other pieces of work that represent your best efforts. (These can be any work from the unit—Problem of the Week, homework, classwork, presentation, and so on.)

# The Role of the POW Project

Discuss how doing *POW 6: Is There Really a Difference?* contributed to the unit for you. You may want to think about two aspects in particular.

- The project's role in helping you *understand* the mathematical ideas

- The project's role in giving you an appreciation for the *usefulness* or *significance* of the mathematical ideas

You can also include comments about your experience of working with a partner on an extended project.

# Appendix

# *Supplemental Problems*

Learning how to compute and interpret the chi-square statistic is the heart of this unit, and many of the supplemental problems are designed to deepen your understanding of this tool and expand your ability to use it. Here are some examples.

- In *Explaining χ² Behavior,* your goal is to explain how the formula for the χ² statistic takes account of variations due to sample size.

- In *Completing the Table,* you'll learn about an important technique for "filling in the gaps" in tables like *A χ² Probability Chart.*

- *Degrees of Freedom* and *A 2 Is a 2 Is a 2, or Is It?* examine how to apply the χ² statistic to situations that are more complex than those you've studied before.

# *Incomplete Reports*

Newspapers, television shows, magazines, journals, and other media often include surveys and statistics in their reports. In these reports, you might get a little information about the statistics or a lot of information. Questions such as, "How many people did you survey?" get answered in some places and not in others.

Your job is to investigate what type of information is given in different surveys. Find examples where practically all the information is revealed, examples where so little information is given that you might question the survey's results, and situations in between.

Some places that you can find such statistics include

- medical journals
- technical documents
- science and health magazines
- news periodicals

1. Comment in detail on at least one of the surveys that you find. Indicate what the strengths and weaknesses of the survey are.

2. If possible, attach copies of surveys that you find.

3. What information should an article present about a survey in order for the results to influence your opinion?

# Two Calls Each

In *POW 4: A Timely Phone Tree*, Leigh and her friends were able to keep making phone calls right up until 9:00. But suppose the parents of all these students decided that no one should make more than two calls a night. How would that affect their phone tree?

In other words, Leigh would still call Mike, from 8:00 until 8:03, and then Leigh and Mike would make their calls to Diane and Ana May from 8:03 to 8:06. But at 8:06, Leigh would be finished, and only the other three would be able to make a call at 8:06. And once Mike made his second call, he'd be finished also.

Under these new rules, how many of Leigh's friends could get her news by 9:00?

You should make the same assumptions as in *POW 4: Timely Phone Tree*.

- It always takes three minutes to make a connection and complete a call.

- No one calls a person who has already been called.

- The caller never gets a busy signal.

# Smokers and Emphysema

Emphysema is a serious lung disease that makes it difficult to breathe.

According to one source, 0.86% of people who smoke develop emphysema, while 0.24% of nonsmokers develop it. (*Note:* Both of these figures are less than 1%.)

Assume for this problem that 15% of adults are smokers. Based on this assumption, answer these questions.

1. What percentage of the population will probably develop emphysema?

2. What percentage of the population will be smokers with emphysema?

3. Given that someone has emphysema, what is the probability that the person is a smoker?

# *Explaining $\chi^2$ Behavior*

You know that for a fair coin, it's much more unusual to get 60% heads out of 100 flips than to get 60% heads out of only 10 flips. In other words, it's "weirder" to get a result of 60 heads and 40 tails than to get a result of 6 heads and 4 tails.

One of the strengths of the $\chi^2$ statistic is that it reflects this difference.

1. To illustrate this aspect of the $\chi^2$ statistic, do these two computations.

   a. Calculate the $\chi^2$ statistic for an experiment in which you get 60 heads in 100 flips.

   b. Calculate the $\chi^2$ statistic for an experiment in which you get 6 heads in 10 flips.

2. Now examine the formula by which the $\chi^2$ statistic is computed and show how this formula leads to getting such different $\chi^2$ statistics in Question 1, even though both experiments had 60% heads.

# Completing the Table

The $\chi^2$ probability chart tells you the probability of getting a $\chi^2$ statistic of a given size or larger when the null hypothesis is true. Unfortunately, the chart only lists some specific values for the $\chi^2$ statistic.

For example, the probability associated with $\chi^2 = 0.6$ is .4386, and the probability associated with $\chi^2 = 0.8$ is .3711. In other words, about 44% of $\chi^2$ statistics are 0.6 or larger, while about 37% are 0.8 or larger. But what can you say about a $\chi^2$ statistic of 0.7? What is the probability of getting a $\chi^2$ statistic of 0.7 or larger?

## The Intuitive Idea

Intuition might lead you to reason something like this:

A $\chi^2$ statistic of 0.7 is exactly halfway between $\chi^2 = 0.6$ and $\chi^2 = 0.8$, so the associated probability should be exactly halfway between .4386 and .3711.

1. Test out this method by following these steps.

   a. Look in the $\chi^2$ probability table to find the probability associated with $\chi^2 = 2.2$ and the probability associated with $\chi^2 = 2.6$.

   b. Find the number that is exactly halfway between the two probabilities that you found in Question 1a.

   c. Compare your answer in Question 1b with the probability in the table for $\chi^2 = 2.4$. Are they exactly the same? Are they close?

## What Actually Happens

What you should have found is that the probability in the table for $\chi^2 = 2.4$ is not exactly halfway—it's actually a little closer to the probability for $\chi^2 = 2.6$ than it is to the probability for $\chi^2 = 2.2$—but it's pretty close to halfway.

*Continued on next page*

It turns out that this method will often give good approximations. The example just described is a special case of a general approximation technique called **linear interpolation.** This technique is especially useful for functions that can't be computed in any simple way, such as the function giving probabilities associated with a given $\chi^2$ statistic.

2. Here's an illustration of this technique with a function that's much simpler than the one associating probabilities with $\chi^2$ statistics. Consider the "squaring function," that is, the function whose equation is $y = x^2$. This can be represented by the notation $f(x) = x^2$.

   a. Find $y$ when $x = 7$. In other words, find $f(7)$.

   b. Find $y$ when $x = 9$. In other words, find $f(9)$.

   Using the technique of a linear interpolation, you would expect $f(8)$ to be halfway between $f(7)$ and $f(9)$.

   c. Check this out. That is, find the number halfway between $f(7)$ and $f(9)$, and then compare it to $f(8)$.

One strength of the technique of linear interpolation is that it isn't limited to "halfway" points.

3. Use the technique of linear interpolation to find the probability associated with a $\chi^2$ statistic of 0.85. (*Hint:* The value 0.85 is one-fourth of the way from 0.8 to 1.0. Find the probability that lies one-fourth of the way from .3711 to .3173. Keep in mind that the probability goes down as the $\chi^2$ statistic goes up.)

4. Develop a general formula for using the technique of linear interpolation, based on this situation:

   Assume that you have a function $g$ and that you know the values of $g(a)$ and $g(b)$. Also assume that $c$ is a number between $a$ and $b$.

   What would you use as an estimate for the value of $g(c)$? (*Hint:* First find a formula in terms of $a$, $b$, and $c$ that tells "what fraction of the way $c$ is from $a$ to $b$." You might want to look at numerical examples for ideas. For example, why is 0.85 "one-fourth of the way" from 0.8 to 1.0?)

# *TV Time*

A researcher wanted to know whether there is a difference in television viewing habits between married and single people.

She worked with a sample of 200 people, of whom 100 were married and 100 were single. Of the married people, 35 were classified as *light* television watchers and 65 were classified as *heavy* watchers. Of the single people, 15 were classified as *light* watchers and 85 were classified as *heavy* watchers.

Do you think the researcher should conclude that there really is a difference between married and single people with respect to their television viewing habits? Explain your answer using the $\chi^2$ statistic.

# *Bigger Tables*

In some problems comparing populations, the data can be described using a 2-by-2 table. But other situations are more complex and require more complicated tables.

In this activity, you will explore what such tables might look like and how they might be used.

1. A newspaper reporter wanted to compare high school students in different grades in terms of their rate of participation in sports. The reporter gathered sample data from several high schools in a major city.

   Before working out all the numbers, the reporter set up this table to reflect the totals of the groups she had surveyed.

|  | 9th graders | 10th graders | 11th graders | 12th graders | Total |
|---|---|---|---|---|---|
| **Plays a sport** |  |  |  |  | 400 |
| **Does not play a sport** |  |  |  |  | 505 |
| **Total** | 250 | 230 | 220 | 205 | 905 |

   If there were no difference from class to class in terms of rate of sports participation, what would be the expected numbers for the individual cells of this table?

2. When the information was tallied, the table turned out to look like this.

|  | 9th graders | 10th graders | 11th graders | 12th graders | Total |
|---|---|---|---|---|---|
| **Plays a sport** | 120 | 110 | 92 | 78 | 400 |
| **Does not play a sport** | 130 | 120 | 128 | 127 | 505 |
| **Total** | 250 | 230 | 220 | 205 | 905 |

   Based on this table, do you think there is a significant difference between the classes in sports participation? If so, how do you think the classes are different? In either case, explain your thinking.

*Continued on next page*

3. The reporter's question was whether different classes had different rates of sports participation. This question required a 2-by-4 table.

   a. Develop five questions that would each require the use of a table that is bigger than 2-by-2. Your questions should deal with situations that are as different as your imagination can make them.

   b. Make up data for two of your situations, like this:

   • For one situation, make up a data set which demonstrates that the underlying populations are significantly different.

   • In the other situation, make up data in which any differences could easily be attributed to sampling variation.

# Degrees of Freedom

You have seen that a 2-by-2 table can be used to compare two populations in terms of whether they possess a given characteristic.

## One Degree of Freedom

Such a table is said to have **one degree of freedom** because if *one* of the cell numbers is known, all the rest can be calculated from it (assuming that you know the row and column totals).

For instance, in the table below concerning school attendance, you are given just one of the four cell values (namely, that fifteen 10th graders were absent).

1. Copy the table below. Then, based on this one cell value (and the row and column totals), fill out the rest of the table. Explain your answers.

|  | 10th graders | 12th graders | Total |
|---|---|---|---|
| **Attended** |  |  | 400 |
| **Absent** | 15 |  | 35 |
| **Total** | 230 | 205 | 435 |

## More Degrees of Freedom

For tables that are larger than 2-by-2, it takes more information to complete the table. The minimum number of cell values needed to complete the table tells you the *degrees of freedom* for the situation.

*Continued on next page*

In this activity, you will investigate what types of tables give different degrees of freedom. The goal is to find a general rule that will tell you the degrees of freedom of an *m*-by-*n* table and explain why the rule works.

2. What is the smallest number of cell values that you would need to know in order to complete the table below? That is, how many degrees of freedom are there in the situation? Explain your answer.

|  | 9th graders | 10th graders | 12th graders | Total |
|---|---|---|---|---|
| **Works after school** |  |  |  | 630 |
| **Does not work after school** |  |  |  | 550 |
| **Total** | 430 | 390 | 360 | 1180 |

3. What is the smallest number of cell values that you would need to know in order to complete the table below? That is, how many degrees of freedom are there in the situation? Explain your answer.

|  | 9th graders | 10th graders | 11th graders | 12th graders | Total |
|---|---|---|---|---|---|
| **Plays a sport** |  |  |  |  | 400 |
| **Does not play a sport** |  |  |  |  | 505 |
| **Total** | 250 | 230 | 220 | 205 | 905 |

4. Look at more tables, make up row and column totals, and see how many degrees of freedom each has. Include examples with more than two rows. (Be sure to check that the sum of the column totals is equal to the sum of the row totals.)

5. Find a rule to calculate the degrees of freedom of an *m*-by-*n* table. Explain why your rule works.

# *A 2 Is a 2 Is a 2, or Is It?*

## *Generalizing the Computation*

The $\chi^2$ statistic is used not only for situations with one degree of freedom but also for tables larger than 2-by-2.

The calculation of the $\chi^2$ statistic for more degrees of freedom is done in essentially the same way as with one degree of freedom:

- First, find the value of the expression $\dfrac{(\text{observed} - \text{expected})^2}{\text{expected}}$ for each cell.

- Then, add these values for all the cells.

The only difference in the computation for a larger table is that there are more cells, and so there are more $\chi^2$ expressions to add.

## *Using the $\chi^2$ Statistic*

In order to use the $\chi^2$ statistic, you need to know which $\chi^2$ statistics would be rare and which would be common if the null hypothesis were true.

In *Is There Really a Difference?* you have investigated the distribution of $\chi^2$ statistics for the case of one degree of freedom. The results are summarized in your $\chi^2$ probability chart. For example, you can look up $\chi^2 = 2.0$ in that chart and see that the probability of getting a $\chi^2$ statistic of 2.0 or larger (when the null hypothesis is true) is approximately .1573.

In this activity, you will compare the distribution of $\chi^2$ statistics for situations with more degrees of freedom to the case of one degree of freedom.

*Continued on next page*

# *A More General χ² Probability Table*

The table at the end of this problem gives $\chi^2$ probabilities for different degrees of freedom. The column headings in the table represent a fixed list of probabilities. Each row of the table is a $\chi^2$ probability chart for a particular number of degrees of freedom.

For example, in the row for one degree of freedom, there is an entry of 1.64. The heading for this column is "0.20" (that is, 20%). This heading tells you that for a situation with one degree of freedom, 20% of all $\chi^2$ statistics will be 1.64 or greater. This is similar to the entry in your $\chi^2$ probability chart, which gives an associated probability of .2059 for a $\chi^2$ statistic of 1.6.

Similarly, in the row for four degrees of freedom, there is an entry of 1.06 under the column heading of "0.90." This means that for a situation with four degrees of freedom, 90% of experiments will give a $\chi^2$ statistic greater than or equal to 1.06.

1. Using this table (and a lot of estimation), give a general description of how the probability of getting a $\chi^2$ statistic of 2.0 or more changes as the number of degrees of freedom goes up.

2. a. Choose a probability "cut-off" for deciding that differences are significant. That is, decide how rare a $\chi^2$ statistic needs to be in order for you to reject the null hypothesis.

   b. Based on your decision in Question 2a, determine, for each degree of freedom in the table, how big the $\chi^2$ statistic needs to be for you to reject the null hypothesis. Explain how you determined that value.

3. Consider the data in this table.

|  | 9th graders | 10th graders | 11th graders | 12th graders | Total |
|---|---|---|---|---|---|
| **Plays a sport** | 120 | 110 | 92 | 78 | 400 |
| **Does not play a sport** | 130 | 120 | 128 | 127 | 505 |
| **Total** | 250 | 230 | 220 | 205 | 905 |

   a. Calculate the $\chi^2$ statistic for this data set.

   b. Based on your result, do you think the different classes participate in sports at the same rate? Explain your answer in terms of the $\chi^2$ statistic and the table.

*Continued on next page*

# *Chi-Square Table for Different Degrees of Freedom*

| Degree of freedom | Probability | | | | | | | | |
|---|---|---|---|---|---|---|---|---|---|
| | .99 | .95 | .90 | .50 | .20 | .10 | .05 | .01 | .001 |
| 1 | 0.00 | 0.00 | 0.02 | 0.45 | 1.64 | 2.71 | 3.84 | 6.63 | 10.83 |
| 2 | 0.02 | 0.10 | 0.21 | 1.39 | 3.22 | 4.61 | 5.99 | 9.21 | 13.82 |
| 3 | 0.11 | 0.35 | 0.58 | 2.37 | 4.64 | 6.25 | 7.81 | 11.34 | 16.27 |
| 4 | 0.30 | 0.71 | 1.06 | 3.36 | 5.99 | 7.78 | 9.49 | 13.28 | 18.47 |
| 5 | 0.55 | 1.15 | 1.61 | 4.35 | 7.29 | 9.24 | 11.07 | 15.09 | 20.52 |
| 6 | 0.87 | 1.64 | 2.20 | 5.35 | 8.56 | 10.64 | 12.59 | 16.81 | 22.46 |
| 7 | 1.24 | 2.17 | 2.83 | 6.35 | 9.80 | 12.02 | 14.07 | 18.48 | 24.32 |
| 8 | 1.65 | 2.73 | 3.49 | 7.34 | 11.03 | 13.36 | 15.51 | 20.09 | 26.12 |
| 9 | 2.09 | 3.33 | 4.17 | 8.34 | 12.24 | 14.68 | 16.92 | 21.67 | 27.88 |
| 10 | 2.56 | 3.94 | 4.87 | 9.34 | 13.44 | 15.99 | 18.31 | 23.21 | 29.59 |

# Do Bees Build It Best?

**Day 1**

# Bees and Containers

Is the honeycomb an efficient shape for storing honey?

This is the central question you will examine in this unit. To start out, you'll make and evaluate some other containers, as you work to define what is meant by an "efficient" container. For your first POW of the unit, you'll be learning more about bees themselves.

*Gabriel Figueroa and Matthew Quesada work together to create a box which will hold the most puffed rice.*

# Building the Biggest

## The Biggest Box

Your main task in this activity is to build the biggest box you can from a single sheet of construction paper. In this activity, "biggest" means "holds the most," and "box" means a container with four rectangular sides, a rectangular bottom, and no top.

You can cut your construction paper and tape pieces together in any way you want as long as your final product is a box. If your first attempt doesn't satisfy you, try again. Keep working at it until you think you have built the biggest box possible.

## Beyond Boxes

Once you have built the biggest *box* you think is possible, try to build a bigger container *of a different shape* (still from a single sheet of construction paper). This time, your shape does not need to have rectangular sides.

# Homework 1      What to Put It In?

All sorts of containers are used for everyday goods. Containers are made out of different materials, have different sizes and shapes, and are used for different purposes. Look around your home for different kinds of containers. If possible, stop at a supermarket or other store for more ideas.

1. What shapes did you find? Make sketches of the different containers you saw.

2. What kinds of materials were used?

3. What units of measurement were used to indicate how much a container holds?

4. What criteria do you think manufacturers use to decide what kind of container to use for their products?

5. Choose a specific container and write about *one* of these two topics:

   • How the container could have been designed better

   • Why the manufacturer designed the container that way

# POW 7 *The Secret Lives of Bees*

The central problem of this unit concerns the honeycombs that bees make for storing honey. This POW is intended to help you understand more about the life of a bee so you have a real-world context for thinking about the central problem.

Your assignment is to write a report on a topic related to bees. Your class should brainstorm to come up with some suggestions.

Your teacher will give you guidelines about how long the report should be and how many sources you should use. Be sure to list the books and magazines you use as well as any other sources of information. You may find it helpful to include drawings. Feel free to express your own opinions and conjectures about why things are the way they are, but be sure to distinguish between your personal opinions and facts obtained from references.

# Days 2-10

# *Area, Geoboards, and Trigonometry*

Measurement is an important theme of this unit.

Once you refine the central question of the unit, you'll be examining how to measure area, because area is one of the fundamental measurement concepts in mathematics.

In developing your understanding of area, you'll move gradually from the concrete examples of figures on the geoboard to the more abstract world of triangles and other polygons. As you will see, right triangles and trigonometry can play a useful role in finding areas.

*Elijah Diaz and Valerie Davila use geoboards to investigate areas of triangles.*

# *Nailing Down Area*

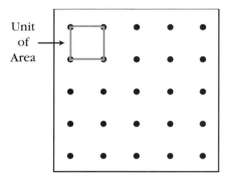

In this activity, the unit of area will be the smallest square on the geoboard, such as the one shown here.

1. Construct each of figures A through L on your geoboard and find their areas. Record your results.

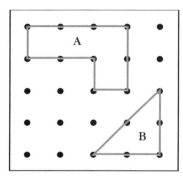

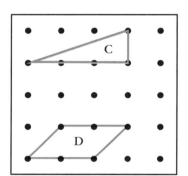

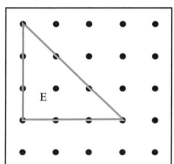

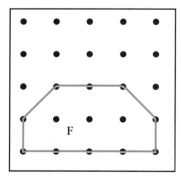

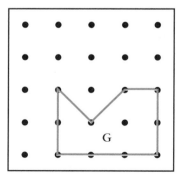

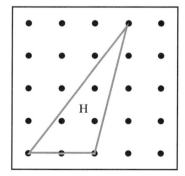

*Continued on next page*

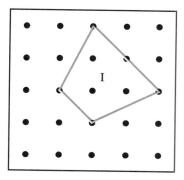 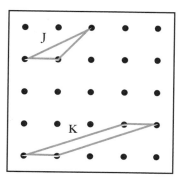 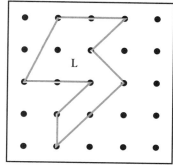

2. Create triangles on the geoboard whose area is half a unit. Find as many different-shaped triangles with this area as you can and record your results on geoboard paper.

3. Create quadrilaterals (four-sided figures) on the geoboard whose area is exactly 1 unit. Find as many different-shaped quadrilaterals with this area as you can and record your results on geoboard paper.

Adapted from *About Teaching Mathematics* by Marilyn Burns, page 99. © 1992 Math Solutions Publications. Used by permission.

# Homework 2                    Approximating Area

For this assignment, you will need to trace this figure onto another sheet of paper and then cut out the shape. You may want to make several copies of this figure.

1. Using the figure you have copied as your unit of area, approximate the area of each of these shapes. You will not be able to find the area for the figure in Question 1b exactly, but do the best you can.

   a.

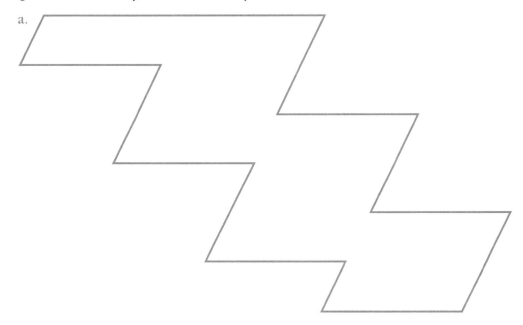

*Continued on next page*

b.

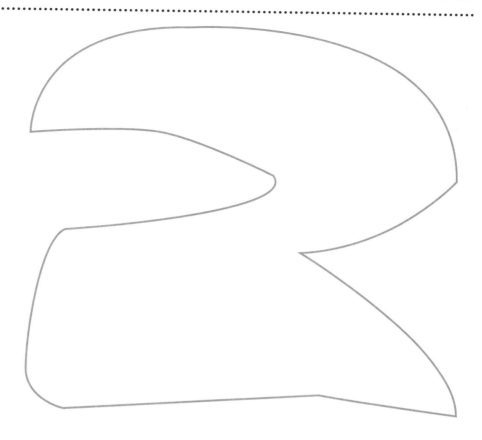

2. Make a new shape that has approximately the same area as the shape in Question 1b.

3. Suppose you had a big pile of pieces of material in the shape and size of the unit you used in Question 1. Estimate how many pieces would be needed to make a medium-size T-shirt and explain how you came up with your estimate.

# Homework 3    How Many Can You Find?

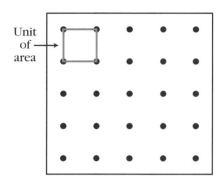

Unit of area →

As in *Nailing Down Area,* the unit of area in this assignment is the basic square on the geoboard, such as the one shown here.

1. Find the area for each of figures A through K.

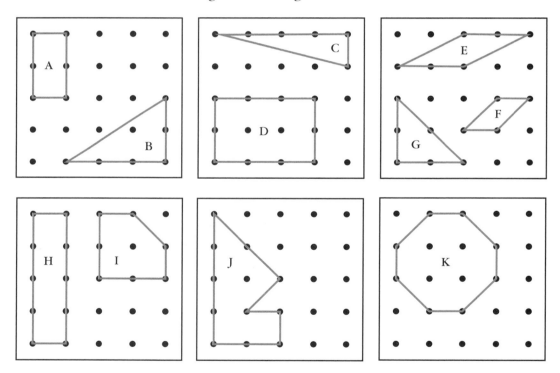

2. Draw polygons on geoboard paper whose areas are 6 units. Find as many different shapes as you can. (You'll probably be better able to find a lot of shapes if you organize your list in some way.)

# That's All There Is!

## Part I: Finding Triangles

In this activity, you will make triangles on the geoboard that fit all of these conditions:

- Each triangle must have an area of 2 units.

- Each triangle must have its vertices on pegs.

- Each triangle must have a horizontal side.

(As in previous activities, the unit of area is the basic square on the geoboard.)

Try to find all the possible shapes for triangles that fit these conditions. As you find such triangles, you should keep track of them so you can include them in the poster you will make in Part II.

## Part II: Studying Triangles

After you have completed Part I, examine the triangles you found and sort them into groups according to one or more of the properties they have in common. Then prepare a poster summarizing your work. Your poster should show and explain your method of organizing the triangles.

## Part III: Further Questions

If you have time, work on these questions.

1. Prove that you found all possible shapes of triangles in Part I.

2. What areas other than 2 units are possible for a triangle that has one horizontal side and has its vertices on pegs?

3. Can you make other triangles with an area of 2 units if you don't require a horizontal side (but still have the vertices on pegs of the geoboard)?

4. Explore Question 3 for triangles with areas other than 2 units.

# Homework 4                            An Area Shortcut?

Your hand is a complicated shape, and finding its area is not an easy task. In this assignment, you'll try it one way and decide whether the "shortcut" a student proposed is a good idea.

1. Trace your hand on a piece of grid paper so that you get a diagram like the one shown here. Find the area of your hand, using one grid square as your unit.

2. When Freddie Short was asked to measure the area of his hand, he proposed this "shortcut."

> I took a piece of string, laid it out around the outline of my hand, and cut off the string so it was exactly the same length as the outline.

> Then I reshaped the string into a more convenient rectangle shape and counted the squares inside the rectangle to get the area of my hand.

Try Freddie's proposed shortcut and decide whether you think his method is a good one or not. Explain your conclusion.

# Homework 5                    Halving Your Way

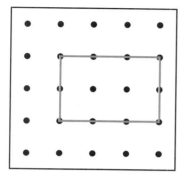

Look at the rectangle shown on the geoboard at the left.

Your task in this assignment is to find ways to cut this rectangle into two pieces with equal area using a rubber band on the geoboard. The rubber band should only be attached at pegs on or inside the rectangle. Your goal is to find all possible ways to do this. (There are exactly 19 ways.) Use geoboard paper to record your work.

The diagrams below show three methods with the dashed line in each diagram representing a rubber band. Although the first two methods are similar, you should count them separately.

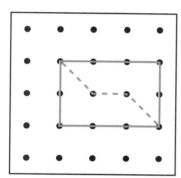 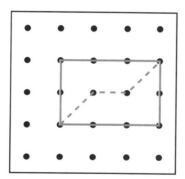 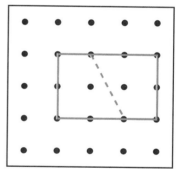

The following method is *not* permitted because it cuts the rectangle into three pieces.

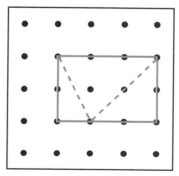

# Homework 6    The Ins and Outs of Area

As you may have realized, the area of a triangle is related to its base and height. Your task in this assignment is to find, and then explain, a rule or formula to describe this relationship.

1. These diagrams show various triangles on geoboards.

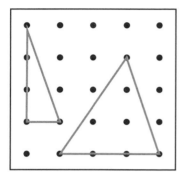 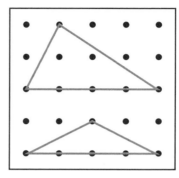 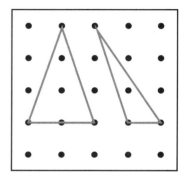

   a. Use an In-Out table like the one below to organize your information about these triangles. Each row of the table should represent one of the triangles.

| In | | Out |
|---|---|---|
| Base of triangle | Height of triangle | Area of triangle |
| | | |

   b. Create more triangles of your own and enter the information in your table.

   c. Find a rule or formula for this table, expressing the area as a function of the base and height. (*Warning:* If you have trouble finding a formula for your table, you may want to check your information, because the formula is not very complicated.)

2. Explain why the formula you found in Question 1b makes sense, using a diagram rather than the pattern from the In-Out table.

# Parallelograms and Trapezoids

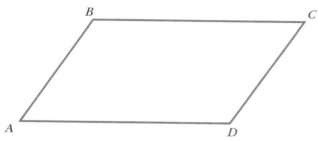

A **parallelogram** is a quadrilateral (a four-sided figure) in which both pairs of opposite sides are parallel. For example, in quadrilateral *ABCD* shown here, $\overline{AB}$ is parallel to $\overline{CD}$ and $\overline{BC}$ is parallel to $\overline{AD}$, so *ABCD* is a parallelogram.

*Note:* A rectangle is a special kind of parallelogram in which the angles are all right angles. For example, in rectangle *KLMN*, $\overline{KL}$ is parallel to $\overline{NM}$ and $\overline{LM}$ is parallel to $\overline{KN}$, so this rectangle is also a parallelogram.

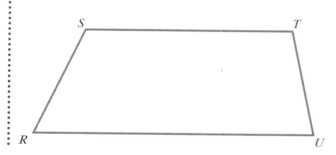

A **trapezoid** is a quadrilateral in which one pair of opposite sides is parallel but the other pair is not. For example, in quadrilateral *RSTU*, $\overline{ST}$ is parallel to $\overline{RU}$ but $\overline{RS}$ is not parallel to $\overline{UT}$, so *RSTU* is a trapezoid.

*Continued on next page*

*Reminder:* Polygons are often named by listing their vertices sequentially. For polygons with more than three vertices, the sequence in which the vertices are named is significant and needs to reflect the way in which the vertices are connected. For instance, trapezoid *RSTU* has other names, such as *TURS* or *SRUT,* but it cannot be called *STRU* because *T* and *R* are not connected by a side of the figure.

Here is a summary of other geometric notation.

$\overleftrightarrow{AB}$       the *line* through two points, *A* and *B*

$\overline{AB}$       the *line segment* from *A* to *B*

$AB$       the *length* of the line segment $\overline{AB}$

$\overrightarrow{AB}$       the *ray* from *A* through *B*

# Going into the Gallery

## The Situation

Yoshi is an emerging young artist who is planning a tour of his work. Because his paintings are quite large, he's concerned about whether they will fit through the doorways of the galleries where they will be displayed.

Each of his paintings is done on a triangular canvas. When they are delivered to galleries, they must be kept upright to avoid damage, with one side of the triangle used as the horizontal base on which the painting will stand as it slides through the doorway.

*Continued on next page*

# *Your Task*

These five triangles show the shapes of some of Yoshi's paintings, using a scale in which 1 centimeter represents 1 meter. For each of these triangles, find the height of the lowest possible doorway through which a painting of that shape will go and state which side (I, II, or III) should be horizontal.

You will be given a copy of these triangles so you can cut them out and hold them upright as if they were sliding through a doorway.

1.

2.

3.

4.

5.

# Homework 7                      Forming Formulas

You have seen that the area of any triangle can be found as half the product of its base and its altitude. For example, in the triangle shown here, the area is $\frac{1}{2}bh$.

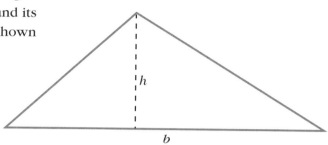

In this assignment, your task is to find two formulas—one for the area of a parallelogram and one for the area of a trapezoid. Part of this task is to decide what measurements are important in determining the area in each case.

Here are two possible approaches to your investigation.

- Use geoboard paper. Draw examples, find their areas, and look for patterns.

- Make parallelograms and trapezoids out of paper. Look for ways you can cut them up and fit the pieces together to make more familiar shapes.

Your write-up for this assignment should include your formulas, some examples for each, and explanations for why your formulas work.

# Homework 8    A Right-Triangle Painting

The shape of a certain right triangle particularly appeals to Yoshi's artistic eye. He has done many paintings using triangular canvases, but his favorites all use right triangles that have an acute angle of 55°, like the triangle shown here. He thinks that perhaps the ratios of the lengths of the sides appeal to him, and he'd like your help in investigating this idea.

1. a. Draw a right triangle *ABC* with a right angle at *C* and a 55° angle at *A*, such as the one shown.

   (*Suggestion:* Start by drawing a 55° angle. Extend both sides of the angle so they are at least 10 centimeters in length. Then draw a right angle from the end of one of the sides, and extend your lines to form a triangle.)

   b. Carefully measure and record the lengths of all three sides. Give your measurements to the nearest millimeter.

2. Find each of these ratios.

   a. $\dfrac{BC}{AB}$

   b. $\dfrac{AC}{AB}$

   c. $\dfrac{BC}{AC}$

   (Recall that *BC* means the *length* of the line segment connecting *B* and *C*.)

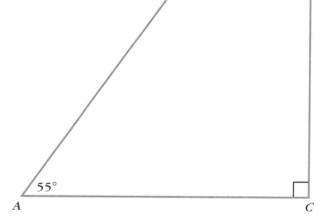

3. Do you think that your classmates will get the same results for Questions 1 and 2 that you got? Explain in detail why or why not.

# POW 8                    *Just Count the Pegs*

Freddie Short has a new shortcut. He has a formula to find the area of any polygon on the geoboard that has no pegs in the interior. His formula is like a rule for an In-Out table in which the *In* is the number of pegs on the boundary and the *Out* is the area of the figure.

Sally Shorter says she has a shortcut for any geoboard polygon with exactly four pegs on the boundary. All you have to tell her is how many pegs it has in the interior, and she can use her formula to find the area immediately.

Frashy Shortest says she has the best formula yet. If you make *any* polygon on the geoboard and tell her both the number of pegs in the interior and the number of pegs on the boundary, her formula will give you the area in a flash!

Your goal in this POW is to find Frashy's "superformula," but you might begin with her friends' more specialized formulas. Here are some suggestions about how to proceed.

*Continued on next page*

1. Begin by trying to find Freddie's formula and some variations, as described in Questions 1a through 1d.

   a. Find a formula for the area of polygons with no pegs in the interior. Your formula should use the number of pegs on the boundary as the *In* and should give you the area as the *Out*. Make specific examples on the geoboard to get data for your table.

   b. Find a different formula that works for polygons with exactly one peg in the interior. Again, use the number of pegs on the boundary as the *In* and the area as the *Out*.

   c. Pick a number bigger than 1, and find a formula for the area of polygons with that number of pegs in the interior.

   d. Do more cases like Question 1c.

2. Find Sally's formula and others like it, as described in Questions 2a through 2c.

   a. Find a formula for the area of polygons with exactly four pegs on the boundary. Your formula should use the number of pegs in the interior as the *In* and should give you the area as the *Out*.

   b. Pick a number other than 4, and find a formula for the area of polygons with that number of pegs on the boundary. Again, use the number of pegs in the interior as the *In* and the area as the *Out*.

   c. Do more cases like Question 2b.

When you have finished work on Questions 1 and 2, look for a superformula that works for all figures. Your formula should have two inputs—the number of pegs in the interior and the number of pegs on the boundary—and the output should be the area of the figure.

Try to be as flashy as Frashy!

# *Write-up*

1. *Problem statement*

2. *Process:* Explain what methods you used to come up with your formulas.

3. *Solution:* Give all the formulas you found.

4. *Evaluation*

5. *Self-assessment*

# A Trigonometric Summary

The concept of similar triangles is the theoretical principle that allows us to define the trigonometric functions as ratios of the sides of right triangles. This statement summarizes the reasoning.

*If an acute angle of one right triangle is the same size as an acute angle of another right triangle, then the triangles are similar, and the ratios of corresponding sides are equal.*

Here's a review of the definitions of the trigonometric functions.

## A Trigonometry Review

Suppose you are given a particular acute angle (in other words, an angle between 0° and 90°). To define the trigonometric functions for that angle, you need to view it as part of a right triangle.

In $\triangle ABC$, $\angle A$ represents the acute angle you are starting with, and $\overline{AC}$ is called the leg **adjacent to** angle $A$, while $\overline{BC}$ is called the leg **opposite** angle $A$.

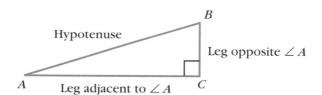

(*Reminder:* $\overline{AC}$ is the *adjacent* leg when you are thinking about $\angle A$, but it would be the *opposite* leg if you were thinking about $\angle B$. Also, recall that in any right triangle, the longest side is called the **hypotenuse** and the two shorter sides are called the **legs**.)

The **sine** of $\angle A$ is the ratio of the length of the leg opposite $\angle A$ to the length of the hypotenuse. The sine of $\angle A$ is abbreviated as **sin** $A$, and the definition is often written simply as

$$\sin A = \frac{\text{opposite}}{\text{hypotenuse}}$$

In $\triangle ABC$,

$$\sin A = \frac{BC}{AB}$$

*Continued on next page*

The **cosine** of $\angle A$ is the ratio of the length of the leg adjacent to $\angle A$ to the length of the hypotenuse. The cosine of $\angle A$ is abbreviated as **cos A,** and the definition is often written simply as

$$\cos A = \frac{\text{adjacent}}{\text{hypotenuse}}$$

In $\triangle ABC$,

$$\cos A = \frac{AC}{AB}$$

The **tangent** of $\angle A$ is the ratio of the length of the leg opposite $\angle A$ to the length of the leg adjacent to $\angle A$. The tangent of $\angle A$ is abbreviated as **tan A,** and the definition is often written as

$$\tan A = \frac{\text{opposite}}{\text{adjacent}}$$

In $\triangle ABC$,

$$\tan A = \frac{BC}{AC}$$

There are three other ratios of side lengths within a right triangle, in addition to those just listed. These ratios, **cotangent, secant,** and **cosecant,** are used less often than those just defined and usually do not have their own calculator keys.

Each is the reciprocal of one of the ratios above, and they can be defined as

$$\text{cotangent } A = \frac{1}{\text{tangent } A}$$

$$\text{secant } A = \frac{1}{\text{cosine } A}$$

$$\text{cosecant } A = \frac{1}{\text{sine } A}$$

They are abbreviated, respectively, as **cot A, sec A,** and **csc A.**

# A Homemade Trig Table

The trigonometric functions are defined as ratios of the lengths of the sides of certain right triangles. One way to understand how these functions work is to draw the triangles, measure the sides, and find the ratios.

In order to get results that are reasonably accurate, you should make all the sides of your triangles at least 5 centimeters long and measure each length to the nearest millimeter.

Your teacher will assign you one or more angles to investigate. For each angle you are assigned, do these three things.

- Draw a right triangle using the assigned angle as one of the acute angles.

- Measure the lengths of all three sides.

- Compute the appropriate ratios to find the sine, cosine, and tangent of the assigned angle.

Compute your ratios to the nearest hundredth.

# Homework 9   More Gallery Measurements

Several friends measured Yoshi's paintings as part of the planning for his tour. They drew little triangles and marked the measurements on each diagram. Unfortunately, they weren't thinking about the doorways and the altitudes, so their measurements weren't necessarily the most useful.

The diagrams in this assignment show the measurements they made. One of the altitudes in each triangle is shown as a dashed line labeled *h*.

Your task in this assignment is to use one of the trigonometric functions—sine, cosine, or tangent—to find the value of *h* in each case.

*Continued on next page*

*Note:* You should base your work on the measurements shown, rather than on any measurements you make yourself.

1.

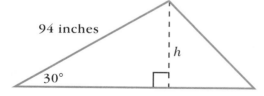

2.

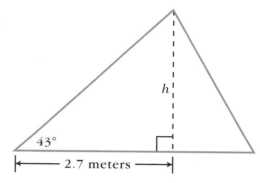

3.

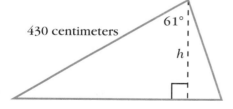

4.

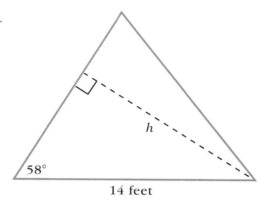

# Homework 10          Shadows and Sailboats

1. A group of students were using shadows to find the heights of trees near their school. They used the diagram shown here to represent the general situation.

   In this diagram, θ (the Greek letter "theta") represents the angle of elevation of the sun, and *S* represents the length of the tree's shadow.

   a. In one case, they found θ = 35° and *S* = 50 feet. What is the height of the tree?

   b. Later that day, with a different tree, they got θ = 60° and *S* = 20 feet. What is the height of that tree?

   c. Develop a general expression for the height of the tree in terms of *S* and θ.

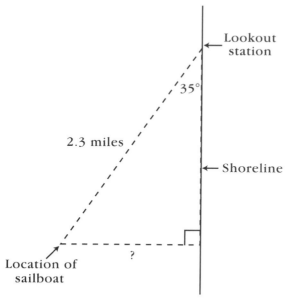

2. a. A sailboat is in trouble, and the people on board are considering trying to swim to shore.

   A lookout station on shore is able to tell them that they are 2.3 miles from the station and that the line from the station to the boat forms an angle of 35° with the shoreline. (Assume that the shoreline is straight, as shown in this diagram.)

   If the people are capable of swimming 1.5 miles, will they be able to make it to shore or should they call for help? Explain.

   b. The lookout station officer would like to be able to tell people in such situations their actual distance from shore. Find a general formula the officer can use to find this distance. Your formula should express this distance in terms of the distance from the station to the boat and the angle between the line to the boat and the shoreline.

**Days 11-16**

*Gabriel Navarro, Gabriel Rodriguez, Jaime Duenas, and Felipe Veloz study the differently shaped triangles in the "Tri-Square Rug" game.*

# A Special Property of Right Triangles

Right triangles are all around us, and the special properties of right triangles have interested people for many centuries.

In the next section of the unit, you'll discover, prove, and apply a property of right triangles that is one of the most famous principles in all of mathematics. This principle was known to civilizations throughout the ancient world, though today it is usually associated with the name of a Greek mathematician.

# *Tri-Square Rug Games*

A rug designer decided to make a rug consisting of three separate square pieces sewn together at their corners, with an empty triangular space between them.

The rug was an immediate hit, and the designer decided to make more of them. He called these creations "tri-square rugs." A sample tri-square rug is shown here.

Al and Betty thought these tri-square rugs could be used to make a great game. They made up these rules.

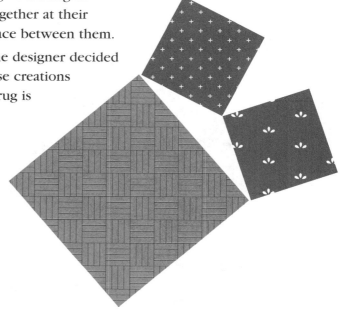

> Let a dart fall randomly on the tri-square rug.
>
> - If it hits the largest of the three squares, Al wins.
>
> - If it hits either of the other two squares, Betty wins.
>
> - If the dart misses the rug, simply let another dart fall.

Your goal in this activity is to decide which tri-square rugs you would prefer if you were Al and which you would prefer if you were Betty, and if there are any rugs that lead to a fair game.

1. You will be given three sheets of chart paper. Label one sheet "Fair Game," a second one "Al Wins in the Long Run," and the third "Betty Wins in the Long Run."

2. Use the squares provided by your teacher to make some sample tri-square rugs. For each tri-square rug you make, decide what would happen in Al and Betty's game in the long run. Then paste the rug carefully on the appropriate sheet of chart paper.

3. When you have several examples for each category, look for a pattern in your results. Your goal is to find a way to tell just by glancing at a tri-square rug who will win in the long run.

# Homework 11 How Big Is It?

1. Summarize what you have learned so far about area, both in terms of what it means and in terms of formulas for computing it.

2. Measurement plays an important role in this unit. For example, you measured the lengths of the sides of right triangles, found areas for figures on the geoboard, and compared volumes for boxes that you built out of construction paper.

   What does it mean to measure length, area, and volume? What are the relationships among the units used for each of these measurements? What do these measurements have in common? How are they different?

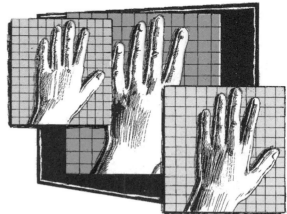

# Any Two Sides Work

The Pythagorean theorem expresses an important relationship between the lengths of the sides of a right triangle.

The problems in this activity give some examples of the many ways this theorem can be used.

1. This diagram shows the longest diagonal on your geoboard.

    a. *Estimate* the length of this diagonal. (As usual, use the distance between adjacent pegs as the unit of length.)

    b. Find the *exact* length of this diagonal using the Pythagorean theorem.

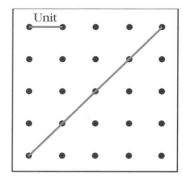

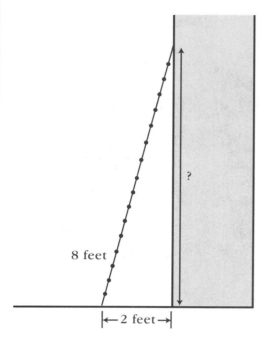

2. An 8-foot ladder is leaning against a wall, as shown in this diagram. The bottom of the ladder is 2 feet from the wall. How high up the wall does the ladder reach?

8 feet

?

|←— 2 feet →|

*Continued on next page*

3. Marlene wants to check that her door frame makes right angles at the corners. The door is 2.5 meters high and 1.5 meters wide. How long should the diagonal of the door be if the corners are right angles?

4. A student thought that the triangle at the right looked like a right triangle, but wasn't sure. Find the length of each side of this triangle, and use your answers to determine *with certainty* whether or not it is a right triangle. Explain your reasoning.

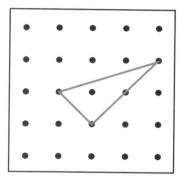

# Homework 12 — Impossible Rugs

Once, when the rug designer was making tri-square rugs, he picked out three square rugs and found that it was impossible to make a tri-square rug from them. He found there was no way to connect them at the corners and leave an empty triangular space in the middle. (Perhaps this happened to you while you worked on *Tri-Square Rug Games*.)

Now the designer wants to know what combinations of squares he *can* use. That is, he wants to know what combinations of squares can be connected at the corners to produce a rug with an empty triangular space in the middle.

1. List some combinations of square rugs that can be used and some combinations that can't be used.

2. Use your examples as a guide to find a rule that will help the designer determine whether three given squares can form a tri-square rug without actually putting them together. Explain your answer.

# Homework 13     Make the Lines Count

There are certainly lots of line segments on the geoboard.

In this assignment, you will investigate the lengths of these segments. For example, the segment labeled *d* in the diagram below is one such segment.

You can find its length by thinking of it as the hypotenuse of a right triangle. Such a triangle can be formed by drawing the segments shown as dashed lines. As usual, the unit of length is the distance between adjacent pegs on the geoboard.

1. Use diagrams to illustrate all the possible lengths for line segments on the geoboard. You should only consider segments that start and end at pegs, and you need to show only one diagram for each length you find.

2. Using the Pythagorean theorem, find each of the lengths that you showed in your answer to Question 1. If you find a length that is not a whole number, you should write it in two ways:

   • As a square root

   • As a decimal rounded to the nearest hundredth

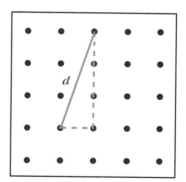

For example, you can find the length *d* in the diagram using the equation $1^2 + 3^2 = d^2$. The length *d* should be given both as $\sqrt{10}$ and as 3.16.

# *Proof by Rugs*

Al and Betty have another game. They began with this right triangle, which has legs of lengths $a$ and $b$ and a hypotenuse of length $c$. Then they made the two square rugs shown below. Each rug has sides of lengths $a + b$, and the triangles within each square are the same as the single right triangle shown at the right.

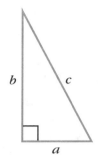

When it's Al's turn, a dart drops on the square rug on the left. If it hits the shaded area, he wins a point. When it's Betty's turn, the dart falls on the square rug on the right. If it hits the shaded area, she wins a point. Assume that the darts always hit the rugs, but that they land randomly within the rug. In other words, all points on a rug have the same chance of being hit.

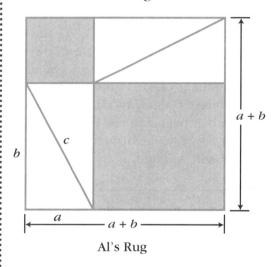

Al's Rug

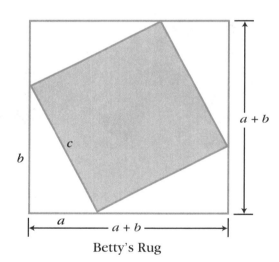

Betty's Rug

1. Is this a fair game? That is, is the chance of the dart landing on the shaded area the same for the two rugs? Explain your answer.

2. How do the two rugs demonstrate that the Pythagorean theorem holds true in general?

# Homework 14    The Power of Pythagoras

As the problems in this assignment show, the Pythagorean theorem can be used to find lengths of many, many kinds of things.

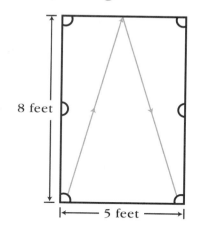

8 feet

5 feet

1. Bonny was doing one of her favorite trick billiard shots. As illustrated in the diagram, her shot started at one corner of the table, hit the exact center of the back cushion, and rebounded into the other corner. How far did her billiard ball travel?

2. The scene is a football field. Ben catches a kick and follows the path shown in the diagram. That is, he starts at the east end of one goal line (near the lower right of the diagram), runs first to the 20-yard line on the west sideline and then to the 50-yard line on the east sideline, where he finally runs for a touchdown at the west end of the other goal line. The field is 53.33 yards wide and 100 yards long. How far does he run altogether?

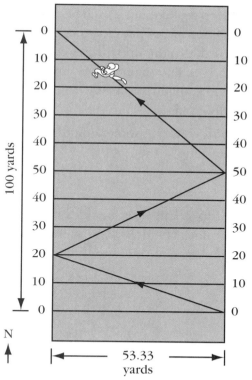

100 yards

N

53.33 yards

3. Corinne and Deanna decide to race from one corner of an open field to the other. The field is rectangular in shape, 60 meters long and 80 meters wide.

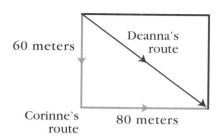

60 meters

Deanna's route

Corinne's route    80 meters

Since Corinne is older and faster, she's going to run along the outside of the field, and Deanna will take the diagonal route. Their paths are indicated by the arrows in the diagram. If Deanna can run at a rate of 5 meters per second, how fast will Corinne have to run to get there at the same time as Deanna?

# POW 9      *Tessellation Pictures*

People around the world have long been fascinated with shapes that can be used for "tessellations." A **tessellation shape** has the property that multiple copies of the shape can be fitted together to meet these conditions.

- No two copies of the shape overlap.
- There are no gaps between copies.
- You can always fit more copies in any direction.

This POW is about tessellation shapes. It has three parts.

## *Part I: Make a Tessellation Shape*

**Step 1:** Take out an index card, and on it, draw some kind of path that goes from the upper-left corner to the upper-right corner. You can use curves, line segments, or any combination, but you should have a path without a "break" or a "loop" in it.

You will have something that looks like the drawing at the right.

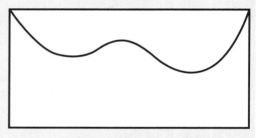

A path has been drawn from upper left to upper right.

**Step 2:** Cut your index card along the curves or lines you drew. This will create two pieces, as shown at the right.

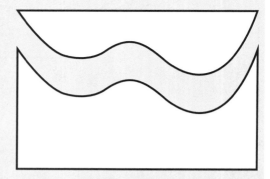

The card has been cut along the path, creating two pieces.

*Continued on next page*

**Step 3:** Tape the two parts together with the "top line" of the upper part of the index card along the bottom of the other piece.

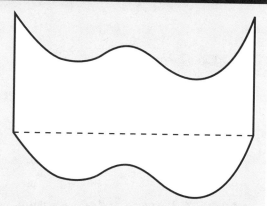

The two pieces of the index card have been taped together, making something more interesting than the rectangle shape.

**Step 4:** Now do a similar thing by drawing a path from what used to be the upper-left corner of the index card to what used to be the lower-left corner. Cut and tape as shown below.

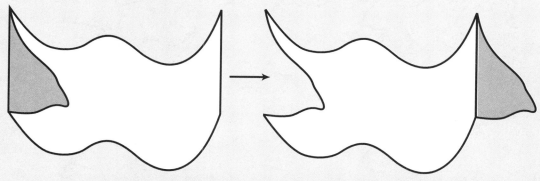

The colored part of the figure on the left has been cut out and taped as shown on the right.

In this example, your final tessellation shape looks like this.

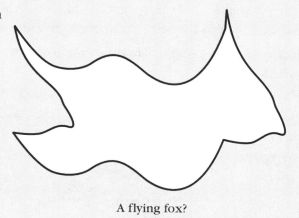

A flying fox?

*Continued on next page*

## Part II: Make Some More Tessellation Shapes

You don't have to start with a rectangle. You can use the method explained in Part I with any parallelogram. (You can also start with any tessellation shape, but the cutting and taping process is more complicated with other starting shapes.)

Make some more tessellation shapes. Don't limit yourself to working with index cards.

## Part III: Make a Tessellation Picture

Choose one of your tessellation shapes. Keep working on it until you get a shape that is really well designed and worth using in your final product.

Now, trace many copies of your shape, fitting them together. For example, if your shape looks like the figure labeled "A flying fox?" shown in Part I, then your drawing might now look like this.

Finally, color in each shape so it looks like something interesting. The different copies of the shape do not all have to be colored the same way.

This product is your tessellation *picture*. It should include at least 10 copies of your tessellation *shape*.

You will turn in this tessellation picture, as well as one copy of each of the other tessellation shapes you made. You will be evaluated primarily on the quality of your tessellation picture, both in terms of its mathematical content and in terms of its beauty.

*Some IMPish Tessellations* shows examples of tessellations that students have made for this POW.

# Some IMPish Tessellations

The tessellation pictures shown here are examples of work that students created in IMP classes at Santa Cruz High School in Santa Cruz, California, and Tamalpais High School in Mill Valley, California.

# Homework 15 Leslie's Fertile Flowers

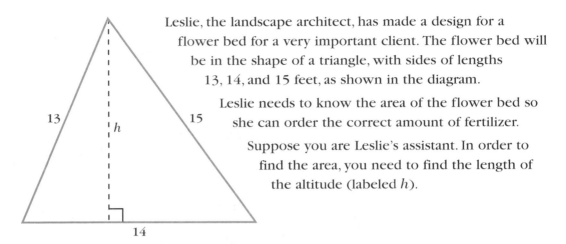

Leslie, the landscape architect, has made a design for a flower bed for a very important client. The flower bed will be in the shape of a triangle, with sides of lengths 13, 14, and 15 feet, as shown in the diagram.

Leslie needs to know the area of the flower bed so she can order the correct amount of fertilizer.

Suppose you are Leslie's assistant. In order to find the area, you need to find the length of the altitude (labeled $h$).

1. Find the length of the altitude. (*Hint:* Guess at ways that the side of length 14 might be split up into two parts, and try to figure out $h$ from that.)

2. Explain how you know that the answer you got in Question 1 is correct. In other words, how are you sure that you made the right guess?

3. While you're thinking about this situation, you might as well calculate the area for Leslie.

# Flowers from Different Sides

After you gave Leslie the area of the flower bed, she suddenly asked, "What if you had looked at the triangle from a different point of view? Would the area still come out the same?" You want to show Leslie how confident you are, so you draw the triangle again, this time using the side of length 15 feet as the base. You use *k* for the length of the new altitude.

1. a. Write two equations using the Pythagorean theorem that you can use to find the value of *k*. (*Hint:* Label the two parts of the new base as *y* and $15 - y$.)

   b. Use your equations to find the value of *k*.

   c. Find the area of the triangle based on this viewpoint.

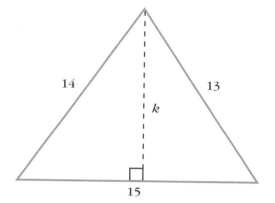

2. Explain to Leslie why you knew all along that the answer would come out the same as before.

3. Find the length of the altitude to the side of length 13, and check the area this way as well.

# Homework 16                    Don't Fence Me In

Rancher Gonzales is building a corral to keep her horses in. She decides that she can afford to buy 300 feet of fencing. For aesthetic reasons, she decides that the corral should be built in the shape of a rectangle. Because she cares about the happiness of her horses, she wants to build the corral so that there is as much space as possible inside it for the horses to move around.

1. Play with the problem, trying various lengths and widths for the rectangle. Use your intuition to guess what the dimensions of the corral should be.

2. Now make an In-Out table where the *In* is the width of the rectangle and the *Out* is the area of the rectangle. Make several specific choices for the width and find the corresponding areas. Pay attention to how you calculate those areas.

3. Use your work in Question 2 to find a formula or rule for this In-Out table. Use the variable $w$ to represent the width of the rectangle and use $A$ to represent the area.

# Days 17-20

# *The Corral Problem*

As you saw in *Homework 16: Don't Fence Me In,* rancher Gonzales is trying to decide what shape to use for her corral. Everybody's got advice for her.

What do you think? Making a good choice will require an understanding of area, perimeter, altitudes, angles, and even trigonometry.

***Cameron Savage and Jessica Sanford put finishing touches on their tessellations POW.***

# *Rectangles Are Boring!*

Rancher Gonzales's nephew Juan has appeared on the scene just as she has decided to build a square corral. Juan thinks this is a totally boring idea. "Why always rectangles?" he asks. "Why not be different? How about a triangle?"

1. If rancher Gonzales uses her 300 feet of fencing to build a corral in the shape of an equilateral triangle, what will the area be?

2. How does that result compare to building a square corral with the 300 feet of fencing?

# Homework 17 More Fencing, Bigger Corrals

Suppose that rancher Gonzales went to the supply store to buy fencing for a square corral and found that they were having a half-price sale. She really loves her horses, so she would spend the same amount she had originally planned. In other words, she would end up purchasing 600 feet of fencing instead of 300, because this would allow her to build a bigger corral.

1. a. What would be the area of rancher Gonzales's square corral if she used 600 feet of fencing?

   b. How does your answer to Question 1a compare to the area of a square corral made from 300 feet of fencing? In other words, what would doubling the perimeter do to the area?

2. What would be the area of an equilateral-triangle corral if rancher Gonzales used 600 feet of fencing? How does that compare to the area of an equilateral-triangle corral made from 300 feet of fencing?

3. What would be the areas of the square and triangular corrals if rancher Gonzales had 900 feet of fencing?

4. What generalizations can you make from your results in Questions 1 through 3? (It might help to make an In-Out table.)

# More Opinions About Corrals

Nephew Juan is not the only one giving rancher Gonzales advice. When friends heard she was building a corral, they all had suggestions to make. When she told them about her experience with triangles and rectangles, they all agreed that she should only consider regular polygons.

One of them thought that a regular pentagon would be a lovely shape for a corral and that this might even give a bigger corral than the square. Your task in this activity is to find the area of a corral that is built in the shape of a regular pentagon and that uses 300 feet of fencing.

# Homework 18              Simply Square Roots

Your work with the Pythagorean theorem has led to many answers involving square roots. In this activity, you will investigate some general questions about the square-root function, and then you will apply your results to specific problems.

You should investigate Questions 1, 2, and 3 by testing specific numbers and then looking for general principles. In other words, choose specific numbers for $a$ and $b$ in each problem, and see whether the property is true or not. For example, in Question 1, you might find the numerical value of both $\sqrt{7 + 10}$ and $\sqrt{7} + \sqrt{10}$ and compare them, and then try other examples.

For each problem, state any general conclusions you reach.

1. Is the square root of a sum equal to the sum of the square roots? In other words, is $\sqrt{a + b}$ the same as $\sqrt{a} + \sqrt{b}$ ?

2. Is the square root of a product equal to the product of the square roots? In other words, is $\sqrt{a \cdot b}$ the same as $\sqrt{a} \cdot \sqrt{b}$ ?

3. Is the square root of a quotient equal to the quotient of the square roots? In other words, is $\sqrt{\dfrac{a}{b}}$ the same as $\dfrac{\sqrt{a}}{\sqrt{b}}$ ?

4. Write each of these square-root expressions in a different way, and explain your answer in terms of your results on Questions 1, 2, and 3.

   a. $\sqrt{25 \cdot 3}$

   b. $\sqrt{49 \cdot 5}$

   c. $\sqrt{18}$

   d. $\sqrt{\dfrac{9}{4}}$

   e. $\sqrt{\dfrac{3}{16}}$

   f. $\sqrt{\dfrac{25}{7}}$

# *Building the Best Fence*

Now rancher Gonzales is really perplexed. She thought squares were good, but now she sees that pentagons are better. She's wondering if maybe there's something even better than pentagons! Based on what she found out about rectangles and triangles and for simplicity and reasons of taste, she decides to consider only regular polygons for the shape of her corral. However, this still leaves a lot of choices.

1. Choose a value greater than 5 for the number of sides. What would the area of the corral be if rancher Gonzales built it in the shape of a regular polygon with this many sides? (Remember, she still has only 300 feet of fencing.)

2. Repeat Question 1 for another regular polygon with more than five sides.

3. Generalize the process used in answering Questions 1 and 2. That is, suppose you have a polygon with $n$ sides (called an ***n*-gon**) with a perimeter of 300 feet. Develop a formula for the area of the corral in terms of $n$.

# Homework 19

# Falling Bridges

Thorough Ted, a construction engineer, correctly computed that the maximum safe load of a bridge being planned would be $1000(99 - 70\sqrt{2})$ tons.

Speedy Sam, the safety supervisor, was asked to design a sign for motorists saying how much weight the bridge could safely hold. He took Thorough Ted's expression, used 1.4 as an approximation for $\sqrt{2}$, and created the sign based on his calculations. But on the day the bridge opened to traffic, it collapsed under a load less than a tenth of the weight shown on Speedy Sam's sign.

Speedy Sam told the city council that he had simply used Ted's figures. Thorough Ted has been over and over his figures and doesn't see how they could be wrong. Write a clear explanation for the city council of why the bridge collapsed.

Adapted from *Mathematics Teacher* (Vol. 82, No. 9, p. 711), copyright December 1989 by the National Council of Teachers of Mathematics.

# Homework 20

# Leslie's Floral Angles

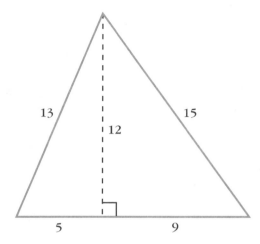

Leslie isn't finished yet with her flower bed from *Homework 15: Leslie's Fertile Flowers.* But thanks to your help, she knows that it should look like this diagram. However, when she finally began to build the flower bed, she realized that she needed to know the angles at the corners.

Can you use some trigonometry to get them? Figure out all the angles to the nearest degree.

**Days
21–27**

# From Two Dimensions to Three

*Molly Bergland and Libby Hobbs discuss
which cylinder in "Which Holds More?" has
the greater volume.*

It's time now to move
from the two-dimensional
world of polygons to the
three-dimensional world of
solid figures. Area continues
to play a role here, because
solid figures have surfaces,
but you'll also be learning
about volume. You will
look at such questions as,
"What is volume?" "How do
you measure volume?" and
"How is volume related to
surface area?"

# Homework 21

# Flat Cubes

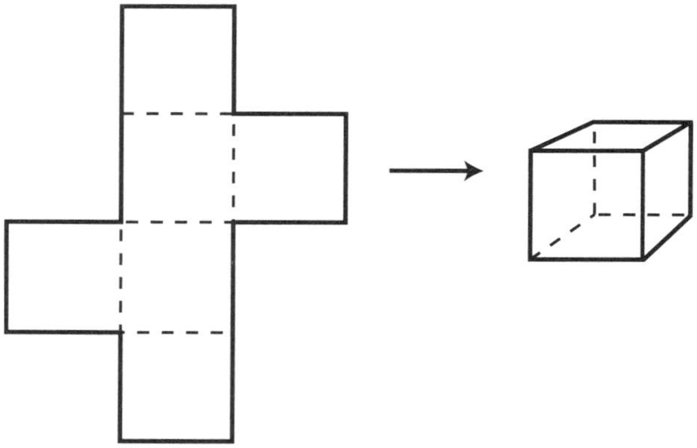

A **net** is a flat pattern that can be cut out and folded to make a solid shape. For instance, the diagram above shows a pattern that can be cut out and folded into the shape of a cube. Thus, the flat pattern is a net for the cube. (You may have made dice from nets like this in the Year 2 unit *Is There Really a Difference?*)

Your assignment is to sketch at least four different outlines that could be cut out and folded into a cube that is 1 inch long in each direction. At least one of your nets should *not* have four squares in a straight line. Use dashed lines to show on your outlines where to make the folds.

# POW 10                    *Possible Patches*

Keisha is making a patchwork quilt. Her quilt will be made up from rectangular patches of material that are each 3 inches by 5 inches.

Rummaging in the attic, she finds a box of old material that her grandmother saved for sewing projects. One item in the box is a rectangular piece of satin that is 17 inches wide and 22 inches long. As you can imagine, Keisha wants to get as many patches as she can from this piece of satin. Each individual patch must be made from a single section of the material. In other words, she won't sew scraps together to make patches.

1. How many 3″-by-5″ patches can Keisha get from her 17″-by-22″ piece of satin? Draw a diagram proving your answer.

2. How many 9″-by-10″ patches could she get from this 17″-by-22″ piece of satin? Can you prove your answer? What about 5″-by-12″ patches? What about 10″-by-12″ patches?

3. Suppose the piece of satin had been 4 inches wide and 18 inches long. Now how many 3″-by-5″ patches would Keisha have been able to get? What if the satin had been 8″-by-9″?

For this POW, begin with the specific situations described in Questions 1 through 3. Then experiment with other patch sizes and other sizes for the piece of satin. Your write-up should give your answers to the specific questions and describe any other results you found.

*Continued on next page*

1. *Problem statement*

2. *Process*

3. *Results:* Include diagrams to justify your answer for the specific questions and to explain any general observations.

4. *Evaluation*

5. *Self-assessment*

Adapted from *Mathematics Teacher* (Vol. 82, No. 8, p. 626), copyright November 1989 by the National Council of Teachers of Mathematics.

# *Flat Boxes*

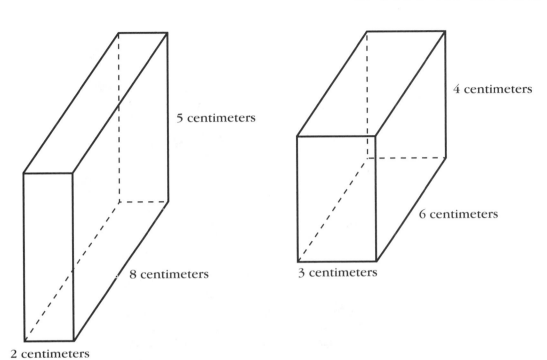

Each of the diagrams above shows a rectangular solid (a box). *Note:* The diagrams are not drawn to scale, so you should use the measurements shown.

1. For each figure, make a net that will fold into the given solid. Your folded net should include both the top and the bottom.

2. Cut out and fold your nets to make each rectangular solid.

3. Find the total surface area for each rectangular solid.

# Homework 22                                     Not a Sound

Pearl was tired of hearing "Turn that music down!" every evening, so she decided to soundproof her room so that she could listen to her music uninterrupted.

Soundproofing material is sold in flat sheets that can be attached to the walls. Pearl decided to put soundproofing on the floor and ceiling too, just to be on the safe side.

1. What if you wanted to soundproof a room where you live? Choose a box-shaped room and figure out how many square feet of soundproofing material you would need to soundproof it. Explain the process you used to figure out your answer.

2. Explain how this problem is related to the unit problem about bees and their honeycombs.

# A Voluminous Task

1. Working with a partner and the cubes provided for you, build each of the ten solid figures shown in this activity. Assume that each top back edge drops straight down so that there are no hidden stacks of cubes.

2. a. Find both the surface area and the volume of each solid figure. Use a single cube as the unit of volume and a face of that cube as the unit of area.

   b. Write a description of how you found the volumes and surface areas. If you used any shortcuts, explain what they were.

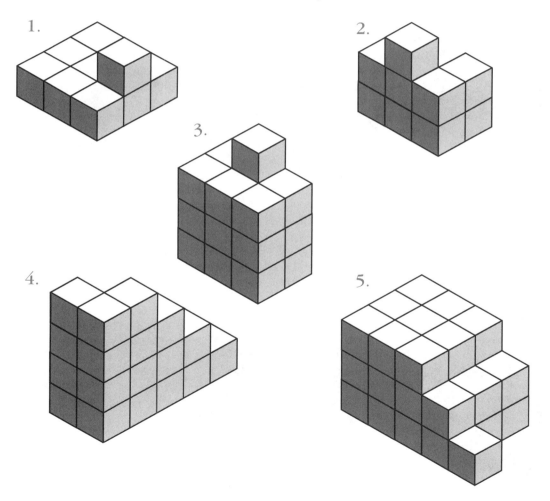

*Continued on next page*

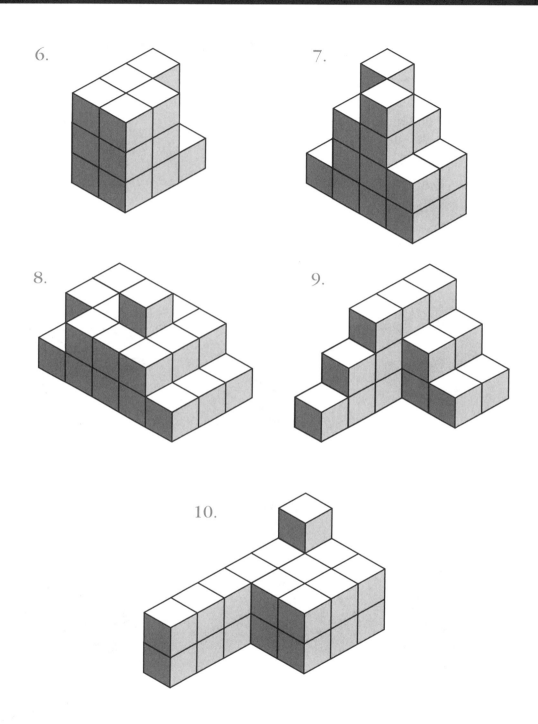

6.

7.

8.

9.

10.

Adapted from E. Ranucci, *Seeing Shapes,* Creative Publications, 1973.

# Homework 23

# Put Your Fist into It

## *Part I: Fists and Volume*

When you make measurements, sometimes you have to improvise. For instance, a gardener may "pace off" the dimensions of a lawn that needs to be seeded, using the length of his or her stride as the unit of length.

In this assignment, you will be using an improvised unit of volume. Although volumes are usually measured with units like cubic centimeters or cubic feet, you will use your fist as the unit.

1. Pick three objects and estimate the volume of each object using your fist as the unit of volume.

2. Estimate the number of cubic inches in your fist.

3. Use your answer to Question 2 to estimate the number of cubic inches in the objects you measured in Question 1.

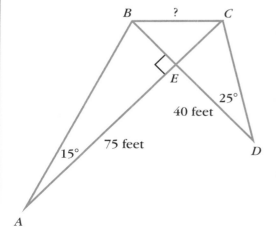

## *Part II: A Triangular Puzzle*

You will need to put various geometric ideas together to solve the puzzle of this diagram. Note that $\overline{BD}$ is perpendicular to $\overline{AC}$, $\angle BAC$ is 15°, $\angle BDC$ is 25°, distance $AE$ is 75 feet, and distance $DE$ is 40 feet.

Find the distance from $B$ to $C$ to the nearest foot, and explain your reasoning.

# The Ins and Outs of Boxes

In *Flat Boxes,* you found the surface areas for two different rectangular solids. To find those surface areas, you needed to know the dimensions of the solids—their lengths, widths, and heights.

In this activity, you will look at how to calculate *both* the surface area *and* the volume of a rectangular solid when you know its length, width, and height. You should use the centimeter as the unit of length, the square centimeter as the unit of surface area, and the cubic centimeter as the unit of volume.

1. Enter the information from *Flat Boxes* in an In-Out table like this one.

| In | | | Out |
|--------|-------|--------|--------------|
| Length | Width | Height | Surface area |
|        |       |        |              |

2. Find the volume for each figure from *Flat Boxes* and enter this information in an In-Out table like the one here.

| In | | | Out |
|--------|-------|--------|--------|
| Length | Width | Height | Volume |
|        |       |        |        |

3. Find rules for each of these two In-Out tables. Use the letters *l, w,* and *h* to represent the length, width, and height, and use *S* and *V* to represent the surface area and volume. Your goal is to express *S* and *V* as functions of *l, w,* and *h.* You may want to examine more rectangular solids in order to get more rows of information for your tables.

   *Note:* The formula for the second table is easier.

4. Justify any formulas you found in Question 3.

# Homework 24    A Sculpture Garden

Penny's neighbor bought a sculpture to put in her garden, and she was really proud of how beautiful it looked. Penny decided that her garden looked drab in comparison, so she decided to build her own outdoor sculpture.

From a previous project she had a pile of wooden crates, all cubes of the same size. Penny chose eight of them that were in good condition. She decided to paint the crate cubes a bright color and pile them on each other. Then her garden would look as nice as her neighbor's.

Penny then realized that she could save money by only painting the parts of the cubes that were exposed. She *did* need to paint the parts of the cubes that would touch the ground to keep them from rotting. She wondered how she could pile the eight cubes so that she would use the least amount of paint.

1. Find a way to arrange Penny's eight cubes that uses the least amount of paint, and draw a picture of this arrangement. Can you find another arrangement that is just as economical?

2. Discuss how the situation in this problem is related to the unit problem about bees and their honeycombs.

# The World of Prisms

You may have used a glass object called a prism as a way to break up light into a rainbow-like pattern. A **prism** is a special type of solid geometric figure, and the object used for breaking up light is a special example of this type of solid.

Geometrically, a prism is formed by moving a plane figure—often a polygon—through space for a fixed distance, keeping it parallel to its original position. The initial and final positions of the plane figure represent the **bases** of the prism. The perpendicular distance between the bases of a prism is called its **height.** Here are some examples of prisms. The shaded face is one of the bases in each case.

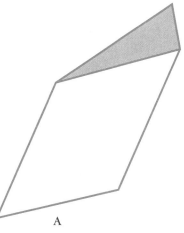

A

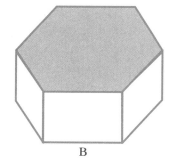

B

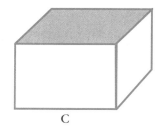

C

Prisms are classified by the shape of their base. Thus, example A is called a **triangular prism;** B is a **hexagonal prism,** and C is a **rectangular prism.**

*Continued on next page*

If the direction through which the base has moved is perpendicular to its original position, the resulting solid is called a **right prism.** An ordinary box is an example of a right rectangular prism. (Examples B and C are intended to show right prisms.) A prism that is not a right prism is called an **oblique prism.** (Example A is intended to represent an oblique prism.)

The faces of a prism other than its bases are called **lateral faces.** (The word "lateral" means *side*.) Because of the way a prism is defined, all of the lateral faces are parallelograms, no matter what the shape of the base. (If the prism is a right prism, the lateral faces are all rectangles.) The sum of the areas of the lateral faces is called the **lateral surface area** of the prism.

The line segments connecting a vertex of the lower base to the corresponding vertex of the upper base is called a **lateral edge.**

# Shedding Light on Prisms

## Part I: Building and Measuring

In this activity, you will build right prisms out of cubes. Examples 1 through 5 below describe different prisms, using a picture or some other information to describe the prism's base. In each case, the base is a polygon.

In each example, find the value of each of these measurements.

    a. Height of the prism

    b. Area of the base

    c. Volume of the prism

    d. Perimeter of the base

    e. Lateral surface area of the prism

You should use the edge of a cube as the unit of length, the face of the cube as the unit of area, and the cube itself as the unit of volume.

Your goal (see Part II) is to find some general relationships among the five measurements listed. You should be thinking about this goal as you work on the individual prisms.

1. A right prism that has the polygon at the right for its base and is four units high

2. A right prism that has the polygon at the right for its base and is six units high

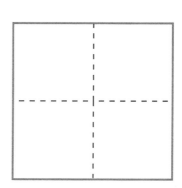

*Continued on next page*

3. A right prism that has the polygon at the right for its base and is three units high

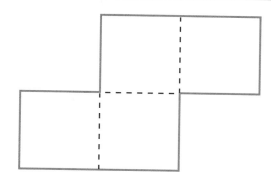

4. A right prism that is nine units high and that has a rectangular base six units in area

5. A right prism that is six units high and that has a nonrectangular base six units in area

# Part II

Study your answers to Part I, and make more prisms if necessary. Find general rules or formulas that show relationships among the measurements you found.

# Homework 25    Pythagoras and the Box

1. Peter bought a very special pen as a birthday present for an artist friend. He has a sturdy box that he wants to use to mail the pen. The box is 4 inches wide, 2 inches deep, and 8 inches high, as shown here.

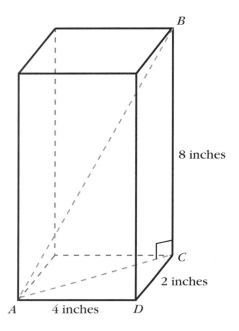

The pen is 10 inches long, so Peter knows he will have to place it in the box along the "long" diagonal—that is, along the line segment in the diagram connecting point *A* to point *B*.

Will the pen fit in this box? (Ignore the thickness of the pen.) Give the length of the long diagonal, and explain how you found the answer.

*Hint:* Begin by finding the length of $\overline{AC}$ using right triangle *ADC*. Then use the fact that $\angle ACB$ is a right angle. You may want to use a shoe box or another box-shaped object to help you picture the situation.

*Continued on next page*

2. a. Find a box-shaped object around the house, and measure its length, width, and height.

   b. Use your measurements from part a to compute the length of the long diagonal of your object.

   c. If possible, confirm your answer to part b by measuring the long diagonal.

3. Generalize your findings from Questions 1 and 2 to a box with dimensions *p, q,* and *r* units. That is, find the length of the segment from point *U* to point *V,* as shown here, and explain your answer.

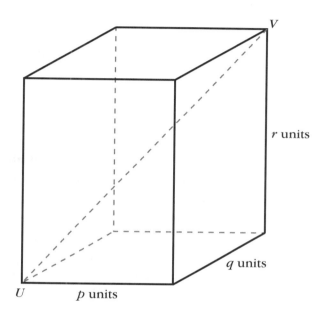

# Homework 26

# Back on the Farm

## 1. A Long Drink

Rancher Gonzales has a neighbor, farmer Minh, who has a drinking trough for his animals. The trough is in the shape of a triangular prism, as shown here. The triangle that forms the base of this prism is a right triangle whose legs are 1 foot long. The trough is 5 feet long.

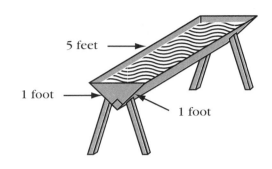

How much water will the trough hold when it is full? Give your answer in cubic feet.

## 2. Farmer Minh's Barn

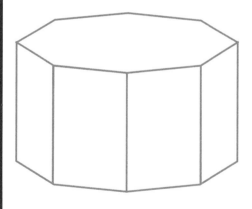

Farmer Minh has been listening to his neighbor discuss corral shapes, and he is thinking about how he can apply her ideas to the barn he's building. The barn will be in the form of a prism with a base that's a regular polygon, something like the prism in the diagram. Of course, the sides will be vertical. He wants the base of the prism to have a perimeter of 300 feet (the same as the perimeter of his neighbor's corral).

Farmer Minh is trying to decide whether to make the barn floor (the prism's base) in the shape of a regular octagon (8 sides), a regular decagon (10 sides), or a regular dodecagon (12 sides). He realizes that he will want to paint the outside of the barn (including the doors), so he decides to look for the shape that gives the barn the least lateral surface area. The barn's walls will be 10 feet tall.

a. What do you think? Before you do any computations, make a guess about which shape for the barn floor will give the least lateral surface area. Write down your guess.

b. Now, make some computations to find the lateral surface area of the barn for each of the three shapes he is considering. Explain your results.

c. How do your results in part b compare to your guess in part a?

# Which Holds More?

Using two identical sheets of binder paper ($8\frac{1}{2}$ inches by 11 inches), make one into a tall, skinny cylinder by taping the two 11-inch sides together where they meet, and make the other into a short, fat cylinder by taping the two $8\frac{1}{2}$-inch sides together where they meet. The base in each case should be a circle. In other words, form two objects like those shown here.

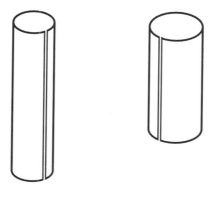

Now, place your cylinders on a flat surface, like a table, so that the flat surface acts as the bottom of the cylinders.

1. Imagine filling them up with something like beans or rice. Guess which one would hold more, or if they would hold the same amount. That is, compare the volumes of the two cylinders. Explain your guess as well as you can.

2. Which of the two cylinders has a greater lateral surface area? Explain your answer.

# Homework 27

# Cereal Box Sizes

A standard-size box of Yummy Crunch cereal is 10 inches tall, 8 inches wide, and 2 inches deep.

1. Find the volume of the box.

2. The manufacturer is thinking about selling Yummy Crunch in other box sizes. Find the volume of each of these sizes.

   a. A "mini-size" that is half as tall, half as wide, and half as deep as the standard size

   b. An "institutional size" that is three times as tall, three times as wide, and three times as deep as the standard size

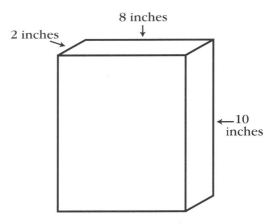

*Continued on next page*

3. Compare the volumes you found in Question 2 with the volume of the standard size. (Look for more detail than simply saying, "The mini-size is smaller and the institutional size is bigger.")

4. a. Find the volume of a "super-size" box that is five times as tall, five times as wide, and five times as deep as the standard size. Try to do this without computing the dimensions of this box. Use your comparison in Question 3 for ideas.

   b. State a general principle for what happens to the volume of a box when each dimension is multiplied by the same amount.

   c. Explain your answer to Question 4b (using diagrams as needed).

5. The manufacturer also wants to know how much cardboard would be used for these various sizes. For the sake of simplicity, ignore both the cardboard wasted in manufacturing the boxes and any places where the cardboard overlaps. In other words, just find their total surface areas.

   a. Find the surface area of the standard-size box.

   b. Find the surface area of the mini-size box.

   c. Find the surface area of the institutional-size box.

6. Compare the surface areas you found in Question 5. (As in Question 3, do more than just say which is bigger and which is smaller.)

7. a. Find the surface area of the super-size box. Try to do this without computing the dimensions of this box.

   b. State a general principle for what happens to the surface area of a box when each dimension is multiplied by the same amount.

   c. Explain your answer to Question 7b (using diagrams as needed).

# Back to the Bees

In the final segment of this unit, you'll combine all the ideas you've been studying with the concept of tessellation to solve the unit problem. You will then compile your portfolios and write your cover letters.

*IMP teacher LeighAnn McCready listens to Nicole Rice summarize the mathematics in the unit.*

# A-Tessellating We Go

The search for the best shape for a honeycomb cell has now been considerably narrowed. You now want to find a shape that meets these conditions.

- The figure is a right prism.
- The base of the prism is a regular polygon.
- The base is a polygon that tessellates.

So you need to answer this question.

*Which regular polygons tessellate?*

Experiment! You can use pattern blocks for some regular polygons. You may want to make your own cut-out versions to investigate the rest.

Once you decide which regular polygons tessellate, look for an explanation that shows that you have found them all.

# Homework 28    A Portfolio of Formulas

In the course of solving the honeycomb problem of *Do Bees Build it Best?* you have developed many formulas related to area and volume. In this assignment, you will collect these formulas in one place so you can include them in your portfolio.

Look back over your work for this unit and write down all the formulas you have used. Write an explanation for each formula, and be sure to define any variables you use. You should include sketches to make the formulas clearer and more useful. Give explanations that will make sense to you when you come back to the portfolio at a future time.

# Homework 29

## *Do Bees Build It Best?* Portfolio

Now that *Do Bees Build It Best?* is completed, it is time to put together your portfolio for the unit. Compiling this portfolio has three parts.

- Writing a cover letter summarizing the unit

- Choosing papers to include from your work in this unit

- Discussing how your understanding of geometry has grown in this unit

## *Cover Letter for "Do Bees Build It Best?"*

Look back over *Do Bees Build It Best?* and describe the central problem of the unit and the main mathematical ideas of the unit. Your description should give an overview of how the key ideas of area and volume were developed and how they were used to solve the central problem.

*Continued on next page*

In compiling your portfolio, you will select some activities that you think were important in developing key ideas of this unit. Your cover letter should include an explanation of why you selected the particular items.

## *Selecting Papers from "Do Bees Build It Best?"*

Your portfolio for *Do Bees Build It Best?* should contain these items.

- *Homework 28: A Portfolio of Formulas*

- A Problem of the Week

   Include either *POW 8: Just Count the Pegs* or *POW 10: Possible Patches*

- Other key activities

   Identify two concepts in this unit for which you think your understanding improved in a significant way. For each concept, choose one or two activities that aided your understanding and explain how each activity helped.

## *Personal Growth*

Your cover letter for *Do Bees Build It Best?* describes how the mathematical ideas developed in this unit. For the final part of your portfolio, write about your own personal development during this unit. You may want to address this question.

   *How do you think you have grown in your understanding of geometry?*

You should include here any other thoughts you might like to share with a reader of your portfolio.

# Appendix

# *Supplemental Problems*

Measurement and the Pythagorean theorem are two of the main mathematical themes of this unit, and they are reflected in the supplemental problems as well. Here are some examples.

- A "measurement medley" presents a series of activities on length, area, and volume.

- *Isosceles Pythagoras* and *Pythagorean Proof* give you further opportunity to see and explain why the Pythagorean theorem is true.

- *Finding the Best Box* and *Another Best Box* follow up on the unit's opening-day activity and ask you to examine the mathematics behind building the "best" box.

# Measurement Medley— Length

Measurement plays an important role in this unit, and length is one of the fundamental aspects of measurement. This activity provides some opportunities for you to work with length.

1. Put a meterstick across the room where you can see it but not reach it.

   a. Use visual estimation to cut pieces of string with each of these lengths.

   • 10 centimeters

   • 38 centimeters

   • 2 meters

   b. Measure your pieces of string and compare the results with what you were trying to get.

2. In this question, you will need the fact that the distance around the earth at the equator is about 40,000 kilometers. Use string and a globe to estimate these distances in kilometers.

   • From San Francisco to New York

   • From Hanoi to Atlanta

   • From Capetown to London

   • Two places of your choice

# *Measurement Medley*
## *—Area*

Area is generally measured in "square units" of some kind. In this activity, you'll work with square centimeters, square kilometers, and square inches.

1. Use centimeter grid paper, a globe, and the fact that the distance around the earth at the equator is about 40,000 kilometers to estimate the area of the continental United States in square kilometers.

2. Suppose that 0.1 ounce of a certain skin cream will cover 50 square inches of skin. Estimate how much cream you would need to cover your arm.

3. Use centimeter grid paper for these tasks and questions.

   a. Draw at least five different rectangles, each with a perimeter of 30 centimeters.

   b. Which of your rectangles in Question 3a has the smallest area? The largest area?

   c. Draw at least five different rectangles, each with an area of 24 square centimeters.

   d. Which of your rectangles in Question 3c has the smallest perimeter? The largest perimeter?

   e. What conclusions can you draw from your work in Questions 3a through 3d?

# *Measurement Medley— Volume*

Length, area, and then comes volume—this activity provides some contexts in which you can explore this third dimension of measurement.

1. Use centimeter cubes and a globe to estimate the volume of the earth in cubic kilometers. Use the fact that the distance around the earth at the equator is about 40,000 kilometers.

2. Find some objects around your home that have approximately these volumes.

   a. 1 cubic inch

   b. 1 cubic foot

   c. 1 cubic meter

3. Fill a jar with marbles and then estimate the fraction of the jar's volume that is air. Do you think this fraction depends on the shape of the jar? Explain.

4. Estimate the volume, in cubic centimeters, of each of these objects.

   a. A penny

   b. A pencil

   c. A rock (find a specific rock to use for this)

# *How Many of Each?*

In this activity, you will use both the triangle and the square from the pattern blocks as units to measure area, and then you will compare the numerical values you get from using different units.

1. Build a shape out of square blocks.

   a. Record the area of your shape using the square as the unit.

   b. Find the approximate area of your shape using the pattern-block triangle as the unit.

2. Build a second shape, using twice as many squares as you used in Question 1.

   a. Record the area of your new shape using the square as the unit.

   b. Find the approximate area of the new shape using the triangle as the unit.

3. a. Draw some figures on paper and find their areas, first using the square as the unit and then using the triangle as the unit.

   b. Record your results (including those from Questions 1 and 2) in an In-Out table, using the area measured in squares as the *In* and the area measured in triangles as the *Out*.

   c. Look for an approximate rule for your table.

# *All Units Are Not Equal*

For this activity, you will need two shapes of different sizes to use as units of area. (You can simply cut two irregular shapes out of stiff paper.)

1. Using the smaller of your shapes as the unit of area, estimate the area of two flat objects around your house. Record those results.

2. Estimate the same areas using your larger shape as the unit, and record those results.

3. Describe how you estimated the areas of these objects. Explain any differences between what you did using the larger unit and what you did using the smaller unit.

4. Imagine that you measured a surface using your smaller shape as the unit and got 5 as the numerical value of the area. What would you get as the area of the same surface if you used your larger shape as the unit? Explain your reasoning.

# Geoboard Squares

In *Checkerboard Squares* (a Problem of the Week from the Year 1 unit *Patterns*), you were asked to find the number of squares (of any size) on an 8-by-8 checkerboard. In that problem, the squares all had vertical and horizontal sides, because they were made of one or more individual colored squares from the checkerboard.

On a geoboard, however, there are other kinds of squares. For example, the geoboard shown here shows a square with slanted sides.

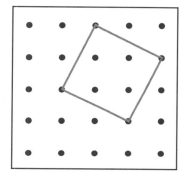

Your task in this activity is to find out how many squares *of all kinds* there are on a geoboard. The only condition is that the vertices of the squares must be at pegs. Don't just find out how many *kinds* of squares there are. Consider every square on the geoboard as an individual example.

*Caution:* Not every four-sided figure is a square. Be sure that all the figures you include are squares.

Start with the standard 5-peg-by-5-peg geoboard, and then consider geoboards of other sizes. Look for a method for calculating the number of squares on an *n*-peg-by-*n*-peg geoboard.

*Extra:* Generalize your results to an *m*-peg-by-*n*-peg rectangular geoboard.

# *More Ways to Halve*

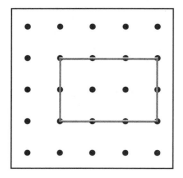

In *Homework 5: Halving Your Way,* you looked at ways to divide the rectangle shown at the left into two parts with equal area. You had to do this with a single rubber band attached at pegs on or inside the rectangle.

In this activity, you can attach the rubber band to any pegs on the geoboard. For example, the diagram at the right shows one new method that you can use. How many ways can you find now?

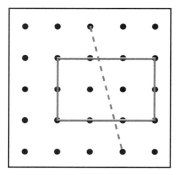

# Isosceles Pythagoras

The Pythagorean theorem says

> **When a triangle has a right angle, the sum of the areas of the squares built on the two legs equals the area of the square built on the hypotenuse.**

Your task in this activity is to find a simple proof of this statement for the special case of an isosceles right triangle.

Consider right triangle *ACB* shown here, where *AC = BC*. Prove that the sum of the areas of the two smaller shaded squares is equal to the area of the larger shaded square. Make your proof as simple as you can.

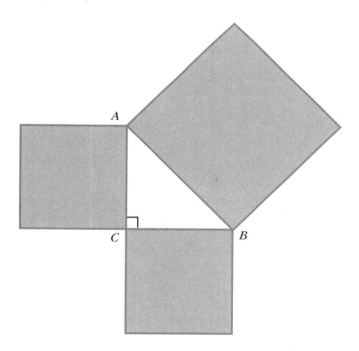

# *More About Pythagorean Rugs*

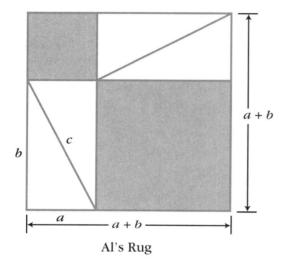

Al's Rug

This diagram was used in *Proof by Rugs* to prove the Pythagorean theorem.

As you saw in that activity, this diagram starts with a right triangle with legs of lengths $a$ and $b$ and a hypotenuse of length $c$. Four copies of that triangle are then placed inside a square of side $a + b$, as shown.

In *Proof by Rugs,* you compared the diagram shown here to a similar one and showed that they had the same amount of shaded area. Your task in this activity is to compare the shaded area in the diagram to the unshaded area *in the same diagram*.

Which is greater—the shaded area or the unshaded area? Does it depend on what triangle you start with? Are there cases where the shaded area and the unshaded area are equal? Prove your answers.

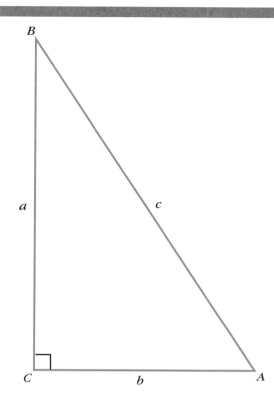

# Pythagorean Proof

Because the Pythagorean theorem is so important, many people have developed different proofs, including people who were not professional mathematicians. For example, James Garfield, who was elected president of the United States in 1880, is credited with creating a proof in 1876 while he served in the House of Representatives.

This activity will help you recreate one of the better-known proofs.

## The Proof

Begin with a general right triangle *ABC,* and use *a, b,* and *c* to represent the lengths of its sides, as shown in the figure above.

Then draw the altitude $\overline{CD}$ from vertex *C* to the hypotenuse, as shown at the right. This creates two smaller right triangles, *ACD* and *BCD,* in addition to the original right triangle.

The proof is based on the fact that the three right triangles are all similar.

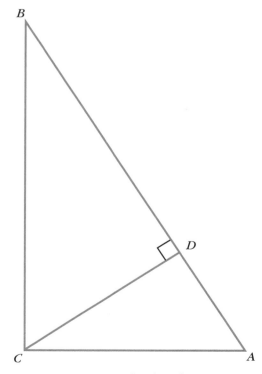

*Continued on next page*

1. Prove that the three triangles *ABC*, *ACD*, and *BCD* are all similar.

The next step of the proof is to set up some proportions based on the similarity. To get this started, notice that $\overline{AB}$, which is the hypotenuse of the large triangle, is broken into two parts by point *D*. Use *x* to represent the length of the segment from *A* to *D* and *c* - *x* to represent the length of the segment from *D* to *B* as shown here.

Based on this labeling, move on to Question 2.

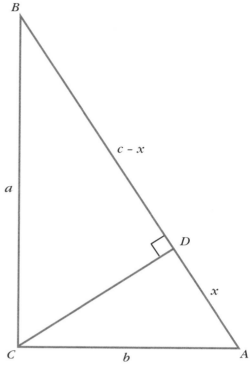

2. Use the fact that the three triangles are similar to write several equations involving *a*, *b*, *c*, and *x*. These equations should state that certain ratios of sides are equal. Here are some hints.

 • Draw diagrams that show each triangle separately and label them. You may want to turn or flip the triangles so that it's clear how the sides match up.

 • Compare each of the two smaller triangles to the large one, rather than compare the two smaller triangles to each other.

3. Work with the equations you developed in Question 2 to show that $a^2 + b^2 = c^2$.

# Hero and His Formula

In *Homework 15: Leslie's Fertile Flowers,* you found the area of a triangle when your only information was the lengths of the three sides. The method used in that problem can be applied to any triangle. In fact, there's a formula known as **Hero's formula** that expresses the area of a triangle in terms of the lengths of its sides.

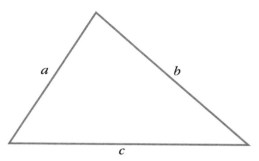

Your task in this activity will be to prove Hero's formula. (*Note:* Questions 2 and 3 are about simplifying the formula shown in Question 1. You should work on Questions 2 and 3 even if you can't prove the formula in Question 1.)

1. Suppose you have a triangle with sides of lengths *a, b,* and *c,* as shown above. Use the method from *Homework 15: Leslie's Fertile Flowers* to show that the area of this triangle is given by the equation

$$A = \frac{1}{4} \sqrt{2a^2b^2 + 2a^2c^2 + 2b^2c^2 - a^4 - b^4 - c^4}$$

*Reminder:* The method in *Homework 15: Leslie's Fertile Flowers* had these five steps.

- Drawing a diagram like this one in which the altitude is shown and where the base is split into two parts of length *x* and *c − x*

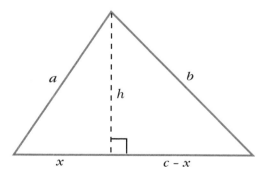

- Using the Pythagorean theorem to write two different equations involving the variables in the diagram

*Continued on next page*

- Combining the two equations to get an equation that does not involve $h$, and solving this equation for $x$

- Finding the height of the triangle

- Finding the area of the triangle

*Note:* When you solve for $x$ (in the third step), you will get an expression involving the three lengths $a$, $b$, and $c$. Finding $h$ once you have $x$ is messy but doesn't involve anything more than the Pythagorean theorem and some algebra.

2. Use the result from Question 1 to show that the area can also be found by the equation

$$A = \frac{1}{4}\sqrt{(a + b + c)(a + b - c)(b + c - a)(a + c - b)}$$

3. The **semiperimeter** of a triangle is defined as half the triangle's perimeter. Hero's formula can be simplified by letting the variable $s$ stand for the semiperimeter. In other words, $s$ is defined by the equation

$$s = \frac{a + b + c}{2}$$

Using this definition for $s$, simplify the result from Question 2 to give the equation

$$A = \sqrt{s(s - a)(s - b)(s - c)}$$

*Note:* This simplification is the usual form of Hero's formula.

# What's the Answer Worth?

Sometimes we measure or estimate something and then make calculations based on these estimates. When your original numbers are only approximations, it's important to know how accurate your final answer is.

1. Two students were studying area. They measured the triangle shown here and found its base to be 4.8 centimeters and its altitude to be 3.4 centimeters. Both measurements were found to the nearest tenth of a centimeter.

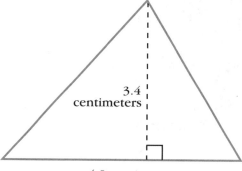

3.4 centimeters

4.8 centimeters

   a. What's the maximum that the base could be? What's the maximum that the altitude could be? (Use the fact that the measurements are rounded to the nearest tenth.)

   b. Use your answers to part a to determine the maximum that the area could be.

   c. What's the minimum that the area could be?

2. Suppose you are taking a test that shows the figure in Question 1 with no explanation and you are asked to find the area of the triangle. How would you answer?

# *Toasts of the Round Table*

Today, when someone proposes a toast, people often clink their glasses together in celebration. Toasting wasn't always done this way. When King Arthur's knights responded to a toast, each knight tapped his lance against the lance of another knight. If possible, every knight tapped lances with every other knight at the table.

As you may know, King Arthur's knights were usually seated at a round table. They didn't want to leave their seats, and the table was large. So, knights opposite each other couldn't always reach far enough to tap lances.

Imagine 30 knights seated at a round table, spaced equally around the edge. Each knight can extend his lance to reach up to 10 feet from his sitting position. (That takes arm length into account also.) The table has a radius of 12 feet.

King Arthur walks into the room and raises his glass. Standing near the table, he drinks in honor of his knights. In response, each knight taps lances with all other knights that are close enough.

1. How many lance taps are there? Explain your answer fully.

2. Consider variations on this problem. Here are some things you might change.

   • The number of knights

   • The radius of the table

   • The distance each knight can reach

Consider changes such as these and generalize as much as you can.

# *All About Escher*

M. C. Escher (1898 – 1972) was a Dutch artist known for his imaginative use of geometry. This picture is taken from one of his creations.

Write a report about Escher, his life, and his work. Find out where he got some of his ideas for designs.

# *Tessellation Variations*

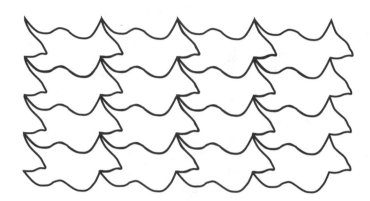

In *POW 9: Tessellation Pictures*, you began with a rectangular index card and did some cutting and pasting to make a new shape. You were then asked to put many copies of your shape together to form a tessellation drawing like the one shown here.

Your task in this activity is to create some new tessellation drawings, this time starting with tessellating figures that are not rectangles. (Triangles and parallelograms are possibilities.)

# The Design of Culture

Geometric patterns, such as tessellations, have been used in the crafts and art of many civilizations as far back as recorded history goes.

Choose a culture—either contemporary or ancient—and describe the use of geometry in some aspect of that culture's artistry. For instance, you might look at weavings, architecture, or fashion.

*Closeup of a Celtic ornamental design from "Celtic Designs," copyright © 1981 by Rebecca McKillip, used by permission of Stemmer House Publishers.*

*Drawing from a Hmong textile created by hill tribe people in Southeast Asia from "Southeast Asian Designs," copyright © 1981 by Caren Caraway, used by permission of Stemmer House Publishers.*

# *Finding the Best Box*

In *Building the Biggest,* you looked at how to build a box with the biggest possible volume from a single sheet of construction paper. As you may recall, you were asked to build a box with four sides and a bottom, but no top.

In this activity, you return to that task, but with a few new details thrown in.

One way to build such a box from a *rectangular* sheet of paper is to begin by cutting out squares at each of the four corners of the sheet. The cuts are shown by the dashed lines in the diagram at the right.

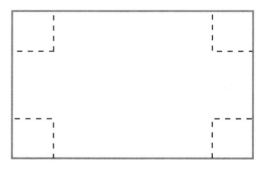

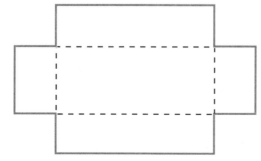

The shape that remains after the corners are cut out is shown in the diagram at the left. This remaining piece of paper can then be folded along the dashed lines, with the "flaps" lifted up together to form the sides of a box. The central rectangular area (within the dashed lines) becomes the bottom of the box.

*Continued on next page*

In this activity, you will focus on what happens if you use this method, starting with a *square* sheet of paper.

1. Start with a sheet of paper that is 12 inches by 12 inches. Use $x$ to represent the side of each cut-out corner.

   a. Find an expression in terms of $x$ for the volume of the resulting box.

   b. Figure out what value of $x$ will maximize this volume.

   c. What is that maximum volume?

2. Next, repeat Questions 1a and 1b, but start with square sheets of paper of different sizes.

3. Describe any patterns you find for how the size of the corner of the "best" box depends on the size of the square sheet of paper.

# Another Best Box

Recycling can change the meaning of "best."

*Finding the Best Box* asks what happens if you build an open-top box by cutting out square corners from a square sheet of paper and folding up the sides. Specifically, Question 1 asks for the maximum possible volume if the sheet of paper is 12 inches by 12 inches.

In that problem, the cut-out corners were basically wasted. Presumably, you could get more volume out of the same sheet of paper if you could somehow use the surface area of those corners. So, suppose you have the full 144 square inches available to build an open-top box. That is, you want to build an open-top box whose total surface area—bottom and four sides—is the same as the total area of a 12-inch–by–12-inch sheet of paper. Assume that the base of your box still has to be a square.

1. Begin with a tall, thin box whose base is a 2-inch–by–2-inch square.

    a. Figure out how tall the box should be in order to get a total surface area of 144 square inches. (Remember that you are counting the bottom and four sides, but no top.)

    b. Find the volume of that box.

2. Now try a short, fat box, and repeat Questions 1a and 1b. (You pick the size of the square base.)

3. To generalize, suppose that the base of the box is a square that is *b* inches on each side. Answer Questions 1a and 1b for this situation. (Your answers will be expressions in terms of *b*.)

4. Find the maximum possible volume and the value of *b* that gives the volume you found in Question 3. (You may need to approximate these answers. You may also want to use a graphing calculator.)

# From Polygons to Prisms

In *Homework 26: Back on the Farm,* you helped farmer Minh minimize the amount of paint he needed for his barn. He is now focusing on maximizing the *volume* of his barn.

Farmer Minh still wants the barn to be in the form of a prism, with vertical sides and a base that's a regular polygon. He has decided that he can afford to buy 3000 square feet of board to use for the lateral sides. He also wants the barn to be 10 feet tall.

What shape should farmer Minh use for the base of the prism to maximize the volume of his barn? Explain your reasoning.

# *What Else Tessellates?*

In *A-Tessellating We Go,* you investigated which regular polygons can be used to tessellate the plane. The diagram at the right shows how equilateral triangles can be used to do this.

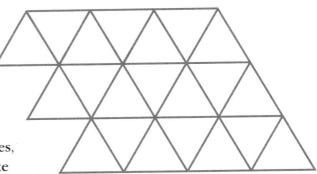

In this activity, you will consider nonregular polygons. Start with triangles, and investigate which of them tessellate and which do not. Then move on to quadrilaterals. See what you can find out and write a report summarizing your findings.

# Cookies

# Cookies and Inequalities

*Emily Perez and Kim Bell discuss homework results within their group.*

The central problem of this unit involves helping a bakery to maximize its profits. The problem is complex. In the opening days of the unit, your main task is to organize all of the information and express the bakery's situation in algebraic terms, using inequalities and linear expressions.

# *How Many of Each Kind?*

Abby and Bing Woo own a small bakery that specializes in cookies. They make only two kinds of cookies—plain and iced. They need to decide *how many dozens* of each kind of cookie to make for tomorrow.

The Woos know that each dozen of their *plain* cookies requires 1 pound of cookie dough (and no icing), and each dozen of their *iced* cookies requires 0.7 pounds of cookie dough and 0.4 pounds of icing. The Woos also know that each dozen of the plain cookies requires about 0.1 hours of preparation time, and each dozen of the iced cookies requires about 0.15 hours of preparation time. Finally, they know that no matter how many of each kind they make, they will be able to sell them all.

The Woos' decision is limited by three factors.

- The ingredients they have on hand—they have 110 pounds of cookie dough and 32 pounds of icing.

- The amount of oven space available—they have room to bake a total of 140 dozen cookies for tomorrow.

- The amount of preparation time available—together they have 15 hours for cookie preparation.

*Continued on next page*

Why on earth should the Woos care how many cookies of each kind they make? Well, you guessed it! They want to make as much profit as possible. The plain cookies sell for $6.00 a dozen and cost $4.50 a dozen to make. The iced cookies sell for $7.00 a dozen and cost $5.00 a dozen to make.

The Big Question is:

*How many dozens of each kind of cookie should Abby and Bing make so that their profit is as high as possible?*

1. a. To begin answering the Big Question, find one combination of dozens of plain cookies and dozens of iced cookies that will satisfy all of the conditions in the problem.

   b. Next, find out how much profit the Woos will make on that combination of cookies.

2. Now find a different combination of dozens of cookies that fits the conditions but that yields a greater profit for the Woos.

This problem was adapted from one in *Introduction to Linear Programming, 2nd Edition,* by R. Stansbury Stockton, Allyn and Bacon, 1963, pp. 19–35.

# Homework 1                    A Simpler Cookie

The Woos have a rather complicated problem to solve. Let's make it simpler. Finding a solution to a simpler problem may lead to a method for solving the original problem.

Assume that the Woos still make both plain and iced cookies, and that they still have 15 hours altogether for cookie preparation. But now assume that they have an unlimited amount of both cookie dough and icing, and that they have an unlimited amount of space in their oven.

The other information is unchanged.

- Preparing a dozen plain cookies requires 0.1 hours.
- Preparing a dozen iced cookies requires 0.15 hours.
- The plain cookies sell for $6.00 a dozen.
- It costs $4.50 a dozen to make plain cookies.
- The iced cookies sell for $7.00 a dozen.
- It costs $5.00 a dozen to make iced cookies.

As before, the Woos know that no matter how many of each kind they make, they will be able to sell them all.

1. Find at least five combinations of plain and iced cookies that the Woos could make without working more than 15 hours. For each combination, find their profit.

2. Find the combination of plain and iced cookies that you think would give the Woos the greatest profit. Explain why you think no other combination will yield a greater profit.

# Homework 2        Investigating Inequalities

## *Part I: Manipulating Inequalities*

In *Solve It!* you used the mystery bags game to think about ways to change equations but keep them true. For instance, if you had a true equation—that is, two expressions that were equal—you could add the same quantity to both sides of the equation, and the resulting expressions would still be equal.

For example, the statement $3 + 8 = 5 + 6$ is true, because $3 + 8$ and $5 + 6$ are both equal to 11. If you add 7 to both sides, the resulting statement is $3 + 8 + 7 = 5 + 6 + 7$ and this statement is also true.

1. The first aspect of Part I is to investigate whether similar principles hold true for inequalities. Start with the inequality $4 > 3$, which is true. For this inequality, perform each of these tasks and then examine whether the resulting statements are true.

   • Add the same number to both sides of the inequality.

   • Subtract the same number from both sides of the inequality.

   • Multiply both sides of the inequality by the same number.

   • Divide both sides of the inequality by the same number.

   For example, if you multiply both sides of the inequality $4 > 3$ by 2, the statement becomes $4 \cdot 2 > 3 \cdot 2$. Your task for each operation is to determine if the new statement is true no matter what "the same number" is.

   Try different possibilities for "the same number," using both positive and negative values.

2. After you finish working with the inequality $4 > 3$, start with a different true inequality and see whether you reach the same conclusions.

3. When you are done exploring, state your conclusions. Make them as general as possible.

*Continued on next page*

## *Part II: Graphing Inequalities*

If an inequality has a single variable in it, we can picture all the numbers that make the inequality true by shading them on a number line. This is called the **graph of the inequality.** An inequality using < or > is called a **strict** inequality. An inequality using ≤ or ≥ is called **nonstrict.**

For example, the colored portion of this number line represents the graph of the strict inequality $x < 4$:

The open circle at the number 4 on the number line means that the number 4 is not included in the graph. (The number 4 is not included because substituting 4 for $x$ gives a false statement.) *Note:* The exclusion of an endpoint is sometimes represented by a parenthesis instead of the open circle.

If we want to include a particular number as part of the graph, we mark that point with a filled-in circle (or by a bracket). For example, the colored portion of the next diagram represents the graph of the nonstrict inequality $x \leq 4$:

4. Draw the graph of the inequality $x > -2$.

5. Draw the graph of the inequality $x \leq 0$.

6. What inequality goes with this graph?

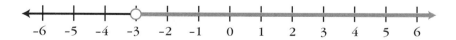

7. How would you use inequalities to describe this graph?

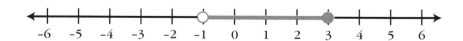

# My Simplest Inequality

In *Homework 2: Investigating Inequalities*, you started with a true inequality involving numbers and explored which operations you could do to both sides that would result in another true inequality.

When inequalities involve variables, we want to know whether the operation produces an **equivalent inequality.** As with equations, two inequalities are called *equivalent* if any number that makes one of them true will also make the other true.

For example, the inequalities $x + 2 < 9$ and $2x + 4 < 18$ are equivalent because, in both cases, the numbers that make them true are precisely the numbers less than 7. For instance, substituting 5 for $x$ makes both true but substituting 10 for $x$ makes both false. (That is, $5 + 2 < 9$ and $2 \cdot 5 + 4 < 18$ are both true, and $10 + 2 < 9$ and $2 \cdot 10 + 4 < 18$ are both false.)

*Continued on next page*

# Part I: One Variable Only

If an inequality has only one variable, you can often find an equivalent inequality that essentially gives the solution. For instance, by subtracting 2 from both sides of the inequality $x + 2 < 9$, you get the equivalent inequality $x < 7$. This tells you that the solutions to $x + 2 < 9$ are the numbers less than 7 (and only those numbers).

1. For each of these inequalities, perform operations to get equivalent inequalities until you obtain one that shows the solution.

   a. $2x + 5 < 8$

   b. $3x - 2 \geq x + 1$

   c. $3x + 7 \leq 5x - 9$

   d. $4 - 2x > 7 + x$

# Part II: Two or More Variables

When an inequality has more than one variable, you can't put it into a form that directly describes the solution. But you can often write the inequality in a simpler equivalent form, such as by combining terms.

For example, suppose you start with the inequality

$$9x - 4y - 2 \geq 3x + 10y + 6$$

You can do these steps to get a sequence of simpler equivalent inequalities.

| | |
|---|---|
| $9x - 2 \geq 3x + 14y + 6$ | (adding $4y$ to both sides) |
| $6x - 2 \geq 14y + 6$ | (subtracting $3x$ from both sides) |
| $6x \geq 14y + 8$ | (adding 2 to both sides) |

Because all the coefficients in the inequality $6x \geq 14y + 8$ are even, you can do the additional step of dividing both sides by 2, to get $3x \geq 7y + 4$. Each of the inequalities in the sequence is equivalent to the original inequality, but $3x \geq 7y + 4$ seems to be the simplest of them all.

2. a. Find numbers for $x$ and $y$ that fit the inequality $3x \geq 7y + 4$.

   b. Substitute the numbers that you found in Question 2a into the original inequality, $9x - 4y - 2 \geq 3x + 10y + 6$, and verify that they make it true.

*Continued on next page*

c. Find numbers for $x$ and $y$ that do not fit the inequality $3x \geq 7y + 4$.

d. Substitute the numbers that you found in Question 2c into the original inequality, $9x - 4y - 2 \geq 3x + 10y + 6$, and verify that they make it false.

e. Explain why steps a through d are not enough to prove that the two inequalities are equivalent.

3. For each of the next three inequalities, perform appropriate operations to get simpler equivalent inequalities.

a. $x + 2y > 3x + y + 2$

b. $\frac{x}{2} - y \leq 3x + 1$

c. $0.2y + 1.4x < 10$

# Homework 3                    Simplifying Cookies

As you have seen, the constraints in the unit problem can be expressed as inequalities using two variables. If you use $P$ to represent the number of dozens of plain cookies and $I$ to represent the number of dozens of iced cookies, one way to write these inequalities is

$$P + 0.7I \leq 110 \qquad \text{(for the amount of cookie dough)}$$

$$0.4I \leq 32 \qquad \text{(for the amount of icing)}$$

$$P + I \leq 140 \qquad \text{(for the amount of oven space)}$$

$$0.1P + 0.15I \leq 15 \qquad \text{(for the amount of the Woos' preparation time)}$$

1. Find at least one equivalent inequality for each of the "cookie inequalities" above. If possible, find an equivalent that you think is simpler than the inequality given.

2. For each of the original inequalities, do each of these steps.

   a. Find a number pair for $P$ and $I$ that fits the inequality and a number pair that does not.

   b. Verify that the number pair that fits the inequality also fits any equivalents you found for that inequality.

   c. Verify that the number pair that does not fit the inequality also does not fit any of the equivalents you found for that inequality.

**Days
4-7**

# *Picturing Cookies*

You have turned the bakery's problem into a set of
inequalities and a profit expression, but that's just a first
step toward understanding the problem. Over the next
several days, you will be examining how to represent
these inequalities using graphs in a way that gives you, at
a glance, a picture of what the Woos' options are.

*Mark Hansen,
Jennifer Rodriguez,
Karla Viramontes, and
Robin LeFevre make
a group graph for a
cookies inequality.*

# *Picturing Cookies—Part I*

By graphing relationships, we can turn symbolic relationships into geometric ones. Because geometric relationships are visual, they are often easier to think about than algebraic statements.

One of the constraints in *How Many of Each Kind?* is that the Woos can make at most 140 dozen cookies altogether (because of oven-space limitations). You can represent this constraint symbolically by the inequality

$$P + I \leq 140$$

where $P$ is the number of dozens of plain cookies and $I$ is the number of dozens of iced cookies.

Choose one color to use for combinations of plain and iced cookies that satisfy the constraint—that is, combinations that total 140 dozen cookies or fewer. Choose a different color for combinations that do not satisfy the constraint—that is, combinations that total more than 140 dozen cookies.

*Continued on next page*

# Some Examples

For instance, what color should you use for the point (20, 50)? In other words, does the combination of 20 dozen plain cookies and 50 dozen iced cookies fit the constraint or not? You can check by substituting 20 for $P$ and 50 for $I$ in the inequality $P + I \leq 140$. Because $20 + 50 \leq 140$ is a true statement, the first color should be used for the point (20, 50).

What about 90 dozen plain cookies and 120 dozen iced cookies? This does not satisfy the constraint, because the statement $90 + 120 \leq 140$ is not true. Therefore, the second color should be used for the point (90, 120).

# Your Task

Your task is to plot both types of points and then to describe the graph of the inequality itself. (The graph of the inequality consists of all points that fit the constraint, that is, all points of the first color.)

Steps 1 through 3 in Question 1 give details on what you need to do. Do your final diagram on a sheet of grid chart paper. If you have time, do Question 2, dealing with other constraints.

1. Go through these steps for the oven-space inequality.

   Step 1: Have each group member try many pairs of numbers for the variables, testing whether each pair satisfies the constraint. *On one shared set of coordinate axes,* group members should plot their number pairs using the appropriate color.

   Step 2: Make sure that your group has many points of both colors. After some experimentation, you may need to change the scale on your axes so that you can show both types of points. If necessary, redraw your axes with a new scale and replot the points you have already found.

   Step 3: Continue with Steps 1 and 2, adding points of each type in the appropriate color. Keep going until you get the "big picture," that is, until you are sure what the overall diagram looks like. Include with your final diagram a statement describing the graph of the inequality itself (the points of the first color) and explaining why you think your description is correct.

2. Graph each of the remaining constraints on its own set of axes. Either follow the process described in steps 1 through 3 of Question 1 or use what you learned in Question 1 about the "big picture."

# Homework 4       Inequality Stories

You have seen that certain real-world situations can be described using inequalities.

For example, in *How Many of Each Kind?* each dozen plain cookies uses 1 pound of cookie dough and each dozen iced cookies uses 0.7 pounds of cookie dough, but the Woos have only 110 pounds of cookie dough. This limitation can be described by the inequality $P + 0.7I \leq 110$, where $P$ is the number of dozens of plain cookies and $I$ is the number of dozens of iced cookies.

In this assignment, you will look further at the relationship between real-world situations and inequalities.

## *Part I: Stories to Inequalities*

In each of Questions 1 and 2, use variables to write an inequality that describes the situation. Be sure to explain what your variables represent.

1. Rancher Gonzales has built the corral for her horses. Now she's building a pen for her pigs. She isn't so worried about efficiency for the pigs, and she decides to go with a boring old rectangle. The pigs need at least 150 square meters of area. She has to decide what dimensions to use for the pen.

2. Al and Betty want to buy a really fancy spinner that costs $200. They each have some money of their own. Al's parents will contribute $2 for every $1 that Al spends. Betty's grandmother will exactly match Betty's contribution. But even if Al and Betty combine all their own money with these additional funds, they still won't have enough.

## *Part II: Inequalities to Stories*

For each of these inequalities, make up a real-world situation that the inequality describes. Again, be sure to explain what your variables represent.

3. $r < t + 2$

4. $a + b + c \leq 30$

5. $x^2 + y^2 \geq 81$ (If you need a hint, think back to some work you did in *Do Bees Build It Best?*)

# Homework 5                          Healthy Animals

Curtis is concerned about the diet he is feeding his pet. A nutritionist has recommended that the pet's diet include at least 30 grams of protein and at least 16 grams of fat per day.

Curtis has two types of foods available—Food A and Food B. Each ounce of Food A supplies 2 grams of protein and 4 grams of fat, while each ounce of Food B supplies 6 grams of protein and 2 grams of fat. Curtis's pet should not eat a total of more than 12 ounces of food per day.

Curtis would like to vary the diet for his pet within these requirements, and so he needs to know what his options are.

1. Choose variables to represent the amount of each type of food Curtis will include in the daily diet. State clearly what the variables represent.

2. Use your variables to write inequalities to describe the constraints of the problem.

3. Choose one of your constraints. Draw a graph that shows which combinations of Food A and Food B satisfy that constraint. Be sure to label your axes and show their scales.

Adapted from *Mathematics With Applications*, by Lial and Miller. ©1987 by Scott, Foresman and Company. Reprinted by permission of Addison-Wesley Educational Publishers Inc.

# POW 11

# *A Hat of a Different Color*

Once upon a time, many years ago and very far away, there lived a wise high school teacher, whose students were always complaining noisily that they had too much homework (and too many POWs, too!).

The wise teacher offered the three noisiest students a deal. He showed them that he had two red hats and three blue hats. The deal worked like this:

> The three students would close their eyes, and while their eyes were closed, the teacher would put a hat on each of their heads (and hide the other two hats).

> Then, one at a time, the students would open their eyes, look at the other two students' heads, and try to determine which color hat was on their own head. At a given student's turn, that student could either guess what color hat he or she had or "pass."

*Continued on next page*

While the first student's eyes were open and that student was still deciding what to do, the other two students kept their eyes closed.

Once the first student either guessed or passed, then the second student could open his or her eyes and either guess or pass. (The eyes of the third student had to remain shut.)

When the second student was finished, then the third student could open his or her eyes and either guess or pass.

Any student who guessed correctly would have no POWs to do the rest of the semester. But any student who guessed wrong would not only have to do the POWs but also help grade everyone else's work. If a student decided to pass, then the work load would stay as usual.

The students drew numbers to see who would go first. Then they closed their eyes, and the wise teacher put a hat on each one's head and hid the remaining two hats.

Arturo, who was first, opened his eyes, looked at the others' heads, and said he wanted to pass. He couldn't tell for sure, and he didn't want to guess in case he was wrong.

Next, Belicia opened her eyes and looked at the others' heads. She also thought about the fact that Arturo had said he couldn't tell. Then she said she didn't want to risk it either. She couldn't tell for sure.

Carletta was third. She just sat there with her eyes still closed tightly and a big grin on her face. "I *know* what color hat I have on," she said. And she gave the right answer.

Your POW is to figure out what color hat Carletta had on and how she knew for sure. The most important part of your POW write-up will be your explanation of how she knew for sure.

*Reminder:* Carletta didn't even look! You should also know that all three students were extremely smart, and if there was a way for them to figure it out, they would be sure to do so.

# *Write-up*

1. *Problem Statement*

2. *Process*

3. *Solution:* Explain how you know for sure what color hat Carletta had on.

4. *Evaluation*

5. *Self-assessment*

# *Picturing Cookies—Part II*

You have already graphed each of the constraints from the unit problem on its own set of axes. Each graph gave you a picture of what that constraint means.

Now you need to see how to combine these constraints to get one picture of all of them together.

1. Begin with one of the constraints that you worked on before. Using a colored pencil, color the set of points that satisfy this constraint. (*Note:* Unlike your work on *Picturing Cookies—Part I,* you should *not* color the points that fail to satisfy the constraint.)

2. Now choose a second constraint from the problem.

   a. On the *same set of axes,* but using a *different color,* color the set of points that satisfy this new constraint.

   b. Using your work so far, identify those points that satisfy *both* your new constraint and the constraint used in Question 1.

3. Continue with the other constraints, using the same set of axes. Use a new color for each new constraint.

   a. Color the set of points that satisfy each new constraint.

   b. After graphing each new constraint, identify those points that satisfy all the constraints graphed so far.

4. When you have finished graphing all the constraints, look at your overall work. Make a single new graph that shows the set of all those points that represent possible combinations of the two types of cookies the Woos can make. In your graph, show all of the lines that come from the constraints, labeled with their equations.

# Homework 6          What's My Inequality?

Graphs of inequalities play an important role in understanding some problem situations. In *Picturing Cookies—Part I,* you started from an algebraic statement—a linear inequality from the unit problem—and saw that its graph was a half plane.

In this assignment, you go from graphs back to algebra. In Part I, you are given graphs that are straight lines, and your task is to find the corresponding linear equations. In Part II, you are given the equation for a straight line, and your task is to find the inequality corresponding to the half plane on one side of that line.

## Part I: Find the Equation

For each of the straight lines in graphs 1 through 4, write a linear equation whose graph is that straight line.

Also describe in words the process by which you found the equation.

1.

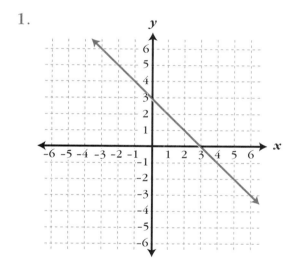

*Continued on next page*

2.

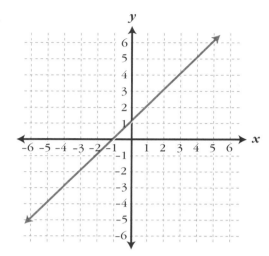

3.

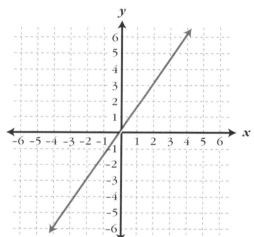

4.

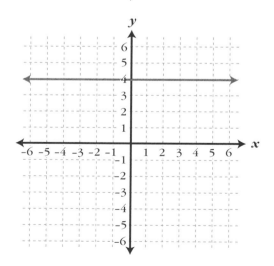

*Continued on next page*

# Part II: Find the Inequality

The shaded area in each of graphs 5 through 8 represents a half plane. (You should imagine that the shaded area continues indefinitely, including all points on the shaded side of the given line.) In each case, you are given an equation for the straight line that forms the boundary of the half plane. Your task is to find a linear inequality whose graph is the half plane itself.

*Note:* If the boundary is shown as a dashed line, it is not considered part of the shaded area. If the boundary is shown as a solid line, it is considered part of the shaded area. This is a common convention, similar to the convention of open and filled-in circles used in *Homework 2: Investigating Inequalities* for graphs in one variable.

5.

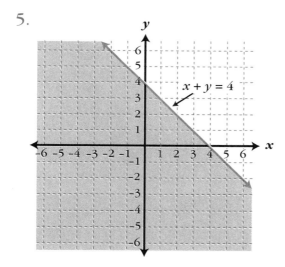

6.

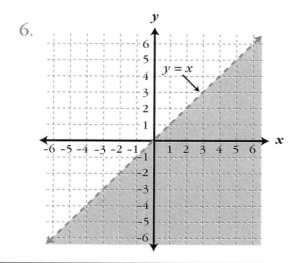

*Continued on next page*

7.

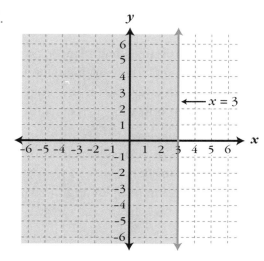

8.

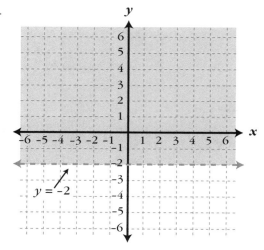

# *Feasible Diets*

You have graphed the individual constraints from *Homework 5: Healthy Animals.* Now your task is to draw the feasible region for that problem.

Here are the key facts.

- Curtis's pet needs at least 30 grams of protein.
- Curtis's pet needs at least 16 grams of fat.
- Each ounce of Food A supplies 2 grams of protein and 4 grams of fat.
- Each ounce of Food B supplies 6 grams of protein and 2 grams of fat.
- Curtis's pet should eat a total of no more than 12 ounces of food per day.

Be sure to identify your variables, label your axes, and show the scales on the axes.

---

# Homework 7                              Picturing Pictures

Hassan is an artist who specializes in geometric designs. He is trying to get ready for a street fair next month.

Hassan paints both watercolors and pastels. Each type of picture takes him about the same amount of time to paint. He figures he has time to do a total of at most 16 pictures.

The materials for each pastel will cost him $5, and the materials for each watercolor will cost him $15. He has $180 to spend on materials. He makes a profit of $40 on each pastel and a profit of $100 on each watercolor.

1. Express Hassan's constraints as inequalities, using $p$ to represent the number of pastels he does and $w$ to represent the number of watercolors.

2. Make a graph that shows Hassan's feasible region. In other words, the graph should show all the combinations of watercolors and pastels that satisfy his constraints.

3. For at least five points on your graph, find the profit that Hassan would make for that combination.

4. Write an algebraic expression to represent Hassan's profit in terms of $p$ and $w$.

# Days 8-14

## *Using the Feasible Region*

As you have seen, the collection of inequalities that describes the central problem of the unit can be represented geometrically as the *feasible region*. But how do you use this region to solve the problem? How do you determine which point in the region will maximize the Woos' profit?

In the next portion of the unit, you will look at several problems similar to the bakery one and examine how geometry can help you find the maximum or minimum value of a linear expression on a feasible region.

*Myrna Yuson, Daniel Escamilla, and Guru Probhakara proudly display their feasible region for "Rock 'n' Rap."*

*Profitable Pictures*

Hassan asked his friend Sharma for advice about what combination of pictures to make. She suggested that he determine a reasonable profit for that month's work and then paint what he needs in order to earn that amount of profit.

Here are the facts you need from *Homework 7: Picturing Pictures.*

- Each pastel requires $5 in materials and earns a profit of $40 for Hassan.
- Each watercolor requires $15 in materials and earns a profit of $100 for Hassan.
- Hassan has $180 to spend on materials.
- Hassan can make at most 16 pictures.

See if you can help Hassan and Sharma. Turn in a written report on the situation. This report should include your work on Questions 1 through 4, but the most important part is your explanation on Question 5.

1. You have already found the feasible region for the problem, which is the set of points that satisfy the constraints. On graph paper, make a copy of this feasible region to use in this problem. Label your axes and show the scales.

2. Suppose Hassan decided $1,000 would be a reasonable profit.

   a. Find three different combinations of watercolors and pastels that would earn Hassan a profit of exactly $1,000.

   b. Mark these three number pairs on your graph from Question 1.

*Continued on next page*

3. Now suppose Hassan wanted to earn only $500 in profit. Find three different combinations of watercolors and pastels that will earn Hassan a profit of exactly $500. *Using a different-colored pencil,* add those points to your graph.

4. Now suppose that Hassan wanted to earn $600 in profit. Find three different combinations of watercolors and pastels that will earn Hassan a profit of exactly $600. *Using a different-colored pencil,* add those points to your graph.

5. Well, Hassan's mother has convinced him that he should try to earn as much as possible. So Hassan needs to figure out the most profit he can earn within his constraints. He also wants to be able to prove to his mother that it is really the maximum amount.

   a. Find the maximum possible profit that Hassan can earn and the combination of pictures he should make to earn that profit.

   b. Write an explanation that would convince Hassan's mother that your answer is correct.

# Homework 8

# Curtis and Hassan Make Choices

1. Curtis goes into the pet store to buy a substantial supply of food for his pet. He sees that Food A cost $2 per pound and that Food B cost $3 per pound. Because he intends to vary his pet's diet from day to day anyway, he isn't especially concerned about how much of each type of food he buys.

a. Suppose that Curtis has $30 to spend. Come up with several combinations of the two foods that he might buy, and plot them on an appropriately labeled graph.

b. Come up with some combinations that Curtis might buy if he were spending $50, and plot them on the same set of axes used in Question 1a.

c. What do you notice about your answers to Questions 1a and 1b?

2. Hassan has a feeling there's going to be a big demand for his work. He is considering changing his prices so that he earns a profit of $50 on each pastel and $175 on each watercolor.

a. Based on these new profits for each type of picture, find some combinations of watercolors and pastels so that Hassan's total profit would be $700, and plot them on a graph. (*Note:* The combinations you give here don't have to fit Hassan's usual constraints.)

b. Now do the same for a total profit of $1,750, using the same set of axes.

c. What do you notice about your answers to Questions 2a and 2b?

# Homework 9     Finding Linear Graphs

Throughout this unit, you are using the graphs of linear equations and inequalities to understand problems. The purpose of this assignment is to have you look at the techniques you use to graph linear equations, and perhaps to find some shortcuts. In class tomorrow, you can share what you have found with others.

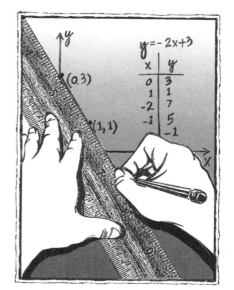

1. One approach to graphing is to make a table of number pairs that fit the equation, graph them, and then connect the points with a straight line.

   a. Create a table of at least five number pairs that satisfy the equation $3x + y = 9$.

   b. Plot the number pairs from your table and connect them with a straight line.

2. Now graph these equations, looking for shortcuts or special techniques. Pay attention to the methods you use, because you will be writing about your methods in Question 3. (*Note:* Read Question 3 before you do Question 2.)

   a. $y = x + 4$

   b. $x + y = 6$

   c. $2x = 3y$

   d. $2x + 3y = 12$

   e. $5y = 6x - 30$

3. Describe in detail what steps you go through when you graph a linear equation. Include any special methods you use that you think might help others. In particular, when you are looking for specific points to plot, how do you decide what numbers to try? If your approach depends on the particular equation, explain how you decide what method to use.

# Hassan's a Hit!

Hassan's pictures are indeed a big hit, especially the watercolors. Based on his success, he is raising his prices as he planned in *Homework 8: Curtis and Hassan Make Choices.* That is, he will now earn a profit of $50 on each pastel and $175 on each watercolor.

Assume that Hassan still has the same constraints. That is, he still has only $180 to spend on materials and can make at most 16 pictures. He had already figured out, with the old prices, how many of each type of picture he should make to maximize his overall profit.

If Hassan wants to maximize his overall profit with the new prices, should he now change the number of pictures he makes of each type? Explain your answer.

# Homework 10     You Are What You Eat

The Hernandez twins do not like breakfast. Given a choice, they would rather skip breakfast and concentrate on lunch.

When pressed, the only things they will eat for breakfast are Sugar Glops and Sweetums cereals. (The twins are allergic to milk, so they eat their cereal dry.)

Mr. Hernandez, on the other hand, thinks his children should eat breakfast every single morning. He also believes that their breakfast should be nutritious. Specifically, he would like them each to get at least 5 grams of protein and not more than 50 grams of carbohydrate each morning.

According to the Sugar Glops package, each ounce of that cereal has 2 grams of protein and 15 grams of carbohydrate. According to the Sweetums box, each ounce of that cereal contains 1 gram of protein and 10 grams of carbohydrate.

So what's the least amount of cereal each twin can eat while satisfying their father's requirements? (Mr. Hernandez wants a proof that his criteria are met, and the twins want a proof that there's no way they can eat less.)

# POW 12                                    *Kick It!*

The Free Thinkers Football League simply has to do things differently. The folks in this league aren't about to score their football games the way everyone else does. So they have thought up this scoring system:

  • Each field goal counts for 5 points.

  • Each touchdown counts for 3 points.

The only way to score points in their league is with field goals or touchdowns or some combination of them.

*Continued on next page*

One of the Free Thinkers has noticed that not every score is possible in their league. For example, a score of 1 point isn't possible, and neither is 2 or 4. But she thinks that beyond a certain number, all scores are possible. In fact, she thinks she knows the highest score that is impossible to make.

1. Figure out what that highest impossible score is for the Free Thinkers Football League. Then explain why you are sure that all higher scores are possible.

2. Make up some other scoring systems (using whole numbers) and see whether there are scores that are impossible to make. Is there always a highest impossible score? If you think so, explain why. If you think there aren't always highest impossible ones, find a rule for when there are and when there are not.

3. In the situations for which there is a highest impossible score, see if you can find any patterns or rules to use to figure out what the highest impossible score is. You may find patterns that apply in some special cases.

# *Write-up*

1. *Problem Statement*

2. *Process:* Include a description of any scoring systems you examined other than the one given in the problem.

3. *Conclusions:*

   a. State what you decided is the highest impossible score for the Free Thinkers' scoring system. Prove both that this score is impossible and that all higher scores are possible.

   b. Describe any results you got for other systems. Include any general ideas or patterns you found that apply to all scoring systems, and prove that they apply in general.

4. *Evaluation*

5. *Self-assessment*

# Homework 11    Changing What You Eat

In *Homework 10: You Are What You Eat,* the twins' solution was simply to eat Sugar Glops. That way, they could get their protein by eating only $2\frac{1}{2}$ ounces of cereal and still not get too many grams of carbohydrate. But what if the cereals had been a little different from the way they were in that problem, or if Mr. Hernandez had been stricter about the twins' carbohydrate intake, or . . . ?

Here are some specific variations for you to work on.

1. Suppose that Sugar Glops is the same as in the original problem (with 2 grams of protein and 15 grams of carbohydrate per ounce), but now Sweetums also has 2 grams of protein per ounce (and still only 10 grams of carbohydrate per ounce). Also suppose that Mr. Hernandez still has a 50-gram limit on carbohydrate and wants each of the twins to get at least 5 grams of protein.

    How much of each cereal should the twins eat if they want to eat as little cereal as possible?

2. Now suppose that Sugar Glops has 3 grams of protein and 20 grams of carbohydrate per ounce, while Sweetums is the same as in Question 1 (2 grams of protein and 10 grams of carbohydrate per ounce). Also suppose that Mr. Hernandez has now decided that the twins can't eat more than 30 grams of carbohydrate (but they still need at least 5 grams of protein).

    What should the twins do?

# Homework 12

# Rock 'n' Rap

The Hits on a Shoestring music company is planning its next month's work. The company makes CDs of both rock and rap music.

It costs the company an average of $15,000 to produce a rock CD and an average of $12,000 to produce a rap CD. (The higher cost for rock comes from needing more instrumentalists for rock CDs.) Also, it takes about 18 hours to produce a rock CD and about 25 hours to produce a rap CD.

The company can afford to spend up to $150,000 on production next month.

Also, according to its agreement with the employee union, the company will spend at least 175 hours on production.

Hits on a Shoestring earns $20,000 in profit on each rock CD it produces and $30,000 in profit on each rap CD it produces. But the company recently promised its distributor that it would not release more rap music than rock, because the distributor thinks the company is more closely associated with rock in the public mind.

The company needs to decide how many of each type of CD to make. *Note:* It can make a fraction of a CD next month and finish it the month after.

1. Graph the feasible region.

2. a. Find at least three combinations of rock and rap CDs that would give the company a profit of $120,000, and mark these points in one color on your graph. (The combinations do not have be in the feasible region.)

   b. In a different color, mark points on your graph that will earn $240,000 in profits.

3. Find out how many CDs the company should make of each type next month to maximize its profit.

4. Explain how you found an answer to Question 3 and why you think your answer gives the maximum profit.

# A Rock 'n' Rap Variation

In *Homework 12: Rock 'n' Rap,* you figured out how many rock CDs and how many rap CDs Hits on a Shoestring should produce to maximize its profit.

Suppose the conditions were the same as in that problem except that the profits were reversed. In other words, suppose the company made $30,000 profit on each rock CD and $20,000 profit on each rap CD.

Would this change your advice to the company about how many CDs of each type to produce to maximize its profit? If so, how many of each type should the company make, and what would be the profit? Explain your answer.

# Homework 13  Getting on Good Terms

Graphing calculators can make it easier to find feasible regions, but in order to draw the graph of an equation on a graphing calculator, the equation needs to be put into "$y =$" form. That is, you need to write the equation so that one variable is expressed in terms of the other. For example, you might rewrite the equation $y - 5 = 4x$ as $y = 4x + 5$.

For each of the equations here, express the variable $y$ in terms of the variable $x$.

1. $y - 2x = 7$
2. $7y = 14x - 21$
3. $5x + 3y = 17$
4. $5(x + 3y) = 2x - 3$
5. $4x - 7y = 2y + 3x$
6. $3y + 7 = 20 - (4x - y)$

# Homework 14      Going Out for Lunch

Imagine! You have just started a full-time summer job in an office. It's your first day on the job, and the boss has sent you out to buy lunch for the 23 people who work in the office.

It turns out that everyone wants either one hot dog or one hamburger for lunch. You get to Enrico's Express, and Enrico asks you how many hot dogs and how many hamburgers you want. You realize you were so excited that you forgot to write down how many of each you were supposed to get. But you see that hamburgers cost $1.50 each and hot dogs cost $1.10 each (tax included), and that you were given $32.10.

## *Your Task*

Assume that the $32.10 is the exact amount needed for your purchase.

1. Figure out, *in any way you can,* how many hot dogs and how many hamburgers were ordered.

2. Do you think the answer you found is the only one possible? Explain why or why not.

Adapted from *Algebra I,* by Paul A. Foerster, Addison-Wesley Publishing Co., 1990, p. 333.

**Days 15–18**

# *Points of Intersection*

The feasible region for a system of inequalities gives you a picture of the possible options, and the family of parallel lines helps you see geometrically where to maximize or minimize a linear expression.

The next step is finding the exact coordinates of that maximum or minimum point. Often, that means finding a common solution for a pair of linear equations. In the main activity for the next few days, *Get the Point,* you will examine pairs of linear equations and develop one or more methods for finding their common solutions.

*Erik Braswell and Casey Kelley will be able to explain several methods for solving systems of equations after meeting the challenge of "Get the Point."*

# Get the Point

In solving problems like the cookie problem, it is helpful to know how to find the coordinates of the point where two lines intersect. As you have seen, this is equivalent to finding the common solution to a system of two linear equations with two variables.

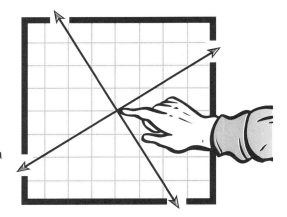

You have probably done this already using either guess-and-check or graphing. Your goal in this activity is to develop an algebraic method, by working with the equations of the two straight lines.

Your written report on this activity should include two things.

- Solutions to Questions 1a through 1e
- The written directions your group develops for Question 2

1. For each of these pairs of equations, find the point of intersection of their graphs by a method other than graphing or guess-and-check. When you think you have each solution, check it by graphing or by substituting the values into the pair of equations.

   a. $y = 3x$ and $y = 2x + 5$

   b. $y = 4x + 5$ and $y = 3x - 7$

   c. $2x + 3y = 13$ and $y = 4x + 1$

   d. $7x - 3y = 31$ and $y - 5 = 3x$

   e. $4x - 3y = -2$ and $2y + 3 = 3x$

2. As a group, develop and write down general directions for finding the coordinates of the point of intersection of two equations for straight lines using an algebraic method, without guessing or graphing. In developing these instructions, you may want to make up some more examples like those in Question 1, either to get ideas or to test whether your instructions work.

   Make your instructions easy to follow so someone else could use them to "get the point."

# Homework 15

# Only One Variable

In *Get the Point,* your goal is to develop a method for solving systems of two linear equations in two variables. In *Solve It!* you saw that you could solve linear equations in one variable, and you may find it helpful to review that process as you work on *Get the Point.*

1. Solve each of these linear equations.

   a. $5x + 7 = 24 - 6x$

   b. $6(x - 2) + 5x = 9x - 2(4 - 3x)$

   c. $\dfrac{x + 3}{2} = 29 - 2x$

   d. $\dfrac{3x + 1}{4} = \dfrac{5 - 2x}{6}$

2. Make up a real-world question that can be represented by a linear equation. Try to create an example in which the variable will appear on both sides of the equation.

# Homework 16         Set It Up

1. You now have more understanding of how to set up and solve pairs of equations in two variables. This problem gives you a chance to apply knowledge.

   Marvelous Marilyn scored 273 points last season for her high school basketball team. Her points resulted from a combination of two-point shots and three-point shots. She made a total of 119 shots. How many of each type of shot did Marilyn make?

   In writing up this problem, follow these steps.

   - Choose variables and state what they represent.

   - Write a pair of equations using your variables that represent the problem.

   - Solve the pair of equations graphically.

   - Solve the pair of equations algebraically.

   - Answer the question in the problem.

   Make up a pair of linear equations whose common solution is
2. $x = 3$ and $y = 5$.

# Homework 17

# A Reflection on Money

1. Read the problem below about Uncle Ralph. You are asked to solve it in two different ways and then reflect on your method of solution.

   Uncle Ralph says that if you can tell him the number of each type of coin in his pocket, then you can have the money. He gives you this information.

   • He has only dimes and quarters.

   • He has 17 coins in his pocket.

   • The coins are worth $3.35.

   Can you get Uncle Ralph's money?

   a. Solve the problem graphically.

   b. Solve the problem algebraically.

   c. Describe the advantages and disadvantages of the graphical and algebraic methods.

2. Here are three systems of linear equations. Solve each system algebraically, using either the method you developed in *Get the Point* or any other methods you have learned. Explain your work clearly.

   a. $y = 2x - 3$ and $3x - 4y = 7$

   b. $c + 2f = -6$ and $3c + f = 2$

   c. $2r - k = -1$ and $6r = 5k - 11$

# POW 13                    *Shuttling Around*

This POW is about solving a puzzle or, rather, about solving a whole set of puzzles. Each of these puzzles requires two sets of markers, such as coins of two different types. We will use plain and shaded circles to represent the markers.

## An Example

One of these puzzles uses three markers of each kind. At the beginning of the puzzle, the markers are arranged as shown below, with each marker in a square. The plain markers are at the left, the shaded markers are at the right, and there is one empty square in the middle.

The task in the puzzle is to move the markers so that the shaded markers end up at the left and the plain markers end up at the right. Of course, there are some rules.

- The plain markers move only to the right and the shaded markers move only to the left.

- A marker can move to an adjacent open square.

- A marker can jump over *one* marker of the other type into an open square.

- No other types of moves are permitted.

## Your Task

The reason that this is a *set* of puzzles is that you can vary the number of markers. Your POW is to investigate this set of puzzles. Begin with the example above, and answer these questions.

1. Can the puzzle be solved? If so, can you find more than one solution?

2. If the puzzle can be solved, how many moves are required? Is there a minimum? Can you prove your answer?

*Continued on next page*

Once you have answered these questions for the case in which there are three markers of each type, look at other examples. Consider only cases in which the numbers of each type are equal and there is exactly one empty square in the middle. (The supplemental problem *Shuttling Variations* examines other cases.)

Here are some things you can do.

- Find out if all such puzzles have solutions, and if so, how many moves are required.

- Look for a rule that describes the minimum number of moves in terms of the number of markers.

- Prove your results.

## Write-up

1. *Problem Statement*

2. *Process:* Describe how you investigated this set of puzzles. Which cases did you examine? Did you try to develop any general conclusions?

3. *Conclusions and Conjectures:* State the questions you investigated and the conclusions you reached. Include questions that you did not have time to investigate. Some of your conclusions may be about specific cases while others may be general.

   If you can prove any of your conclusions, include the proofs. If any of your conclusions are still tentative, label them as conjectures.

4. *Evaluation*

5. *Self-assessment*

# Homework 18                                    More Linear Systems

This assignment continues the work with linear equations. Keep alert for new shortcuts and new insights into how to solve either one-variable or two-variable equations.

1. Find the solution to each of these linear equations.

   a. $3(c + 4) - 2c = 16 - 4(c + 5)$

   b. $t + 2(t - 4) = 5(1 - 2t)$

   c. $\dfrac{r + 5}{2} = 12 - 3r$

   d. $7w + 2(3 - 2w) = 4(w + 2) - (w - 6)$

2. Find the value of both variables in each of these linear systems.

   a. $4a - 5b = -4$ and $3a + 6b = 10$

   b. $u - v = 3$ and $2u + 2v = 10$

   c. $2x + 3y = 1$ and $6y = 7 - 4x$

$t + 2(t - 4) = 5(1 - 2t)$

$\frac{r+5}{2} = 12 - 3r$

$4a - 5b = -4$

$u - v = 3$

Linear
Equations
Hall

# Days 19–21

## Cookies and the University

You're now ready to solve the unit problem. Use the feasible region, a family of parallel profit lines, and a pair of linear equations to find the cookie combination that the Woos are looking for.

When you're done with that, you'll apply the ideas to a completely new problem about college admissions.

*Vanessa Morales, Mandy Mazik, Anandika Muni, and Moses Lazo work together to solve the unit problem.*

# "How Many of Each Kind?" Revisited

As you saw in *How Many of Each Kind?* Abby and Bing Woo have a small bakery that makes two kinds of cookies—plain cookies and cookies with icing. Here is a summary of the key information about the Woos' situation.

## Summary of the Situation

Facts about making the cookies:

- Each dozen plain cookies requires 1 pound of cookie dough.
- Each dozen iced cookies requires 0.7 pounds of cookie dough and 0.4 pounds of icing.
- Each dozen plain cookies requires 0.1 hours of preparation time.
- Each dozen iced cookies requires 0.15 hours of preparation time.

Constraints:

- The Woos have 110 pounds of cookie dough and 32 pounds of icing.
- The Woos have room to bake a total of 140 dozen cookies.
- The Woos together have 15 hours available for cookie preparation.

Costs, prices, and sales:

- Plain cookies cost $4.50 a dozen to make and sell for $6.00 a dozen.
- Iced cookies cost $5.00 a dozen to make and sell for $7.00 a dozen.
- No matter how many of each kind they make, the Woos will be able to sell them all.

The Big Question is:

*How many dozens of each kind of cookie should Abby and Bing make so that their profit is as high as possible?*

*Continued on next page*

# *Your Assignment*

Imagine that your group is a business consulting team, and the Woos have come to you for help. Of course, you want to give them the right answer. But you also want to explain to them clearly how you know that you have the best possible answer so that they will consult your group in the future.

You may want to review what you already know from earlier work on this problem. Look at your notes and earlier assignments. Then write a report for the Woos. Your report should cover these items.

- An answer to the Woos' dilemma, including a summary of how much cookie dough, icing, and preparation time they will use, and how many dozen cookies they will make altogether

- An explanation for the Woos that will convince them that your answer gives them the most profit

- Any graphs, charts, equations, or diagrams that are needed as part of your explanation

You should write your report based on the assumption that the Woos do not know the techniques you have learned in this unit about solving this type of problem.

# Homework 19                           A Charity Rock

## Part I: Solving Systems

Solve each pair of equations.

1. $5x + 2y = 11$ and $x + y = 4$

2. $2p + 5q = 15$ and $6p + 15q = -29$

3. $3a + b = 4$ and $6a + 2b = 8$

## Part II: Rocking Pebbles

At concerts given by the group Rocking Pebbles, some of the tickets sold are for reserved seats and the rest are general admission.

For a recent series of two weekend concerts, the Pebbles pledged to give their favorite charity an amount equal to half of what was paid for

general-admission tickets. After the concerts, the charity called the Pebbles' manager to find out how much money the charity would be getting.

The manager looked up the records. She found that for the first night, 230 reserved-seat tickets and 835 general-admission tickets were sold. For the second night, 250 reserved-seat tickets and 980 general-admission tickets were sold.

The manager saw that the total amount of money collected for tickets was $23,600 for the first night and $27,100 for the second night, but she didn't know the prices for the two different kinds of tickets. (The prices were the same at both concerts.)

Figure out what the two ticket prices were, and use that information to tell the manager how much the Pebbles will give to the charity. As part of your work on this problem, set up a pair of linear equations. Then solve this system in whatever way you like, such as using algebra, graphs, or guess-and-check.

# Homework 20                           Back on the Trail

The two problems here are similar to ones that you saw in the *Overland Trail* unit. But now you know more about writing and solving equations, so these should be easier than when you first saw them.

For each of these two problems, write a pair of equations using two variables, and then solve the equations to answer the question.

## Part I: Fair Share on Chores

A family with three boys and two girls needs to split up the chore of watching the animals. Altogether, the animals need to be watched for 10 hours.

If the length of each boy's shift is an hour more than the length of each girl's shift, how long is each type of shift? (Remember that the family considered this fair in light of other chores the boys and girls had to do.)

## Part II: Water Rationing

The Stevens family contains three adults and five children. The Muster family contains two adults and four children. In a typical day on the trail, the Stevenses use about 15 gallons of water (for drinking, washing, and so on), while the Musters use about 11 gallons per day.

Assuming that each adult uses about the same amount and that each child uses about the same amount, how much does each use in a typical day?

# *Big State U*

The Admissions Office at Big State University needs to decide how many in-state students and how many out-of-state students to admit to the next class. Like many universities, Big State U has limited resources, and budget considerations have to play a part in admissions policy.

Here are the constraints on the Admissions Office decision.

- The college president wants this class to contribute a total of at least $2,500,000 to the school after it graduates. In the past, Big State U has gotten an average of $8,000 in contributions from each in-state student admitted and an average of $2,000 from each out-of-state student admitted.

- The faculty at the college wants entering students with good grade-point averages. Grades of in-state students average less than grades of out-of-state students. Therefore, the faculty is urging the school to admit at least as many out-of-state students as in-state students.

- The housing office is not able to spend more than $85,000 to cover costs such as meals and utilities for students in dormitories during vacation periods. Because out-of-state students are more likely to stay on campus during vacations, the housing office needs to take these differences into account. In-state students will cost the office an average of $100 each for vacation-time expenses, while out-of-state students will cost an average of $200 each.

The college treasurer needs to minimize educational costs. Because students take different courses, it costs an average of $7,200 a year to teach an in-state student and an average of $6,000 a year to teach an out-of-state student.

Your job is to recommend how many students from each category should be admitted to Big State U. You need to minimize educational costs, as the treasurer requires, within the constraints set by the college president, the faculty, and the housing office.

Your write-up should include a proof that your solution is the best possible within the constraints. Show any graphs that seem helpful, and explain your reasoning carefully.

Adapted from *An Introduction to Mathematical Models in the Social and Life Sciences,* by Michael Olinick, Addison-Wesley, 1978, p. 169.

# Homework 21      Inventing Problems

You have now seen several problems that you could solve by defining variables and then setting up and solving a pair of linear equations. Examples include *Homework 14: Going Out for Lunch, Homework 19: A Charity Rock,* and *Homework 20: Back on the Trail.*

In this assignment, you get to make up your own problem.

1. Make up a problem that you think can be solved with two equations and two unknowns.

2. Solve the problem and write up your solution. (*Note:* As you work on the problem, you may find that you want to change it in some way to improve it.)

3. Write out your problem (without solution) on a separate sheet of paper. Put your name on this sheet. (Tomorrow, students will share and work on one another's problems.)

# Days 22-27

## *Creating Problems*

Over the course of this unit, you have solved a variety of problems involving linear equations, linear inequalities, and graphs. One way to get more insight into such problems is to create one of your own. In *Homework 21: Inventing Problems,* you created a two-equation/two-unknown problem. In the final days of the unit, you will create a linear programming problem.

*Ryan Jones, Keith Landrum, and Mandy Ledford prepare the graph for the linear programming problem they created.*

# Homework 22

# Ideas for Linear Programming Problems

In tomorrow's activity, *Producing Programming Problems,* your group will be creating a linear programming problem to solve and present to the class.

In any linear programming problem, there are variables, constraints, and something to minimize or maximize. For instance, the central problem of this unit (*How Many of Each Kind?*) is a linear programming problem. The variables represented the number of dozen plain cookies and the number of dozen iced cookies. The constraints were the Woos' available preparation time and oven space and the amounts of cookie mix and icing mix. The goal was to maximize the profit.

*Big State U* is another linear programming problem. In that problem, the task was to minimize the educational costs, using variables representing the number of in-state students and the number of out-of-state students to be admitted. The constraints involved contributions to the university, grade-point averages, and housing costs.

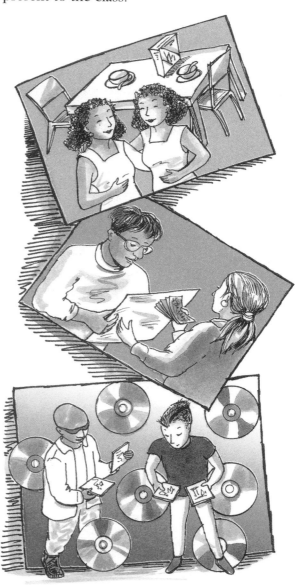

1. Reread *Homework 10: You Are What You Eat.*

   a. What did the variables in that problem represent?

   b. What were the constraints?

   c. What needed to be maximized or minimized?

*Continued on next page*

2. Reread the activity *Profitable Pictures.*

   a. What did the variables in that problem represent?

   b. What were the constraints?

   c. What needed to be maximized or minimized?

3. Reread *Homework 12: Rock 'n' Rap.*

   a. What did the variables in that problem represent?

   b. What were the constraints?

   c. What needed to be maximized or minimized?

4. Create a situation in which you might be interested in maximizing or minimizing something. Describe the situation, and state what you would maximize or minimize.

5. Choose two variables to go with your situation from Question 4, and give two or three constraints using those variables that might apply.

# *Producing Programming Problems*

Your group is to make up a linear programming problem. Here are the key ingredients you need to have in your problem.

- Two variables
- Something to be maximized or minimized that is a *linear function* of those variables
- Three or four *linear* constraints

Once you have written your problem, you must solve it.

Then you should put together an interesting 5-to 10-minute presentation. This presentation should do three things.

- Explain the problem
- Provide a solution to the problem
- Prove that there is no better solution

# Homework 23

# Beginning Portfolio Selection

The main problem for this unit, *How Many of Each Kind?* is an example of a **linear programming** problem. You have seen several such problems, including *Profitable Pictures*, *Homework 10: You Are What You Eat*, *Homework 12: Rock 'n' Rap*, and *Big State U.*

1. Describe the steps you must go through to solve such a problem.

2. Pick three activities from the unit that helped you to understand particular steps of this process, and explain how each activity helped you. (You do not need to restrict yourself to the activities listed above.)

   *Note:* Selecting these activities and writing the accompanying explanations are the first steps toward compiling your portfolio for this unit.

# Homework 24     Just for Curiosity's Sake

## Part I: Solving Equations

Solve each pair of equations.

1. $3s + t = 13$ and $2s - 4t = 18$
2. $6(a + 2) - b = 31$ and $5a - 2(b - 3) = 23$
3. $z - w = 6$ and $5z + 3w = 10$

## Part II: Rocking Pebbles

The Rocking Pebbles just finished another two-night series of concerts in a different town. This time, none of the shows was for charity, because the producer in that town thought doing so might set a bad precedent. The Pebbles' manager wonders how the producer priced the tickets. (The prices may have been different from those in the previous town but they were the same both nights.)

The producer said there were 200 reserved-seat tickets and 800 general-admission tickets sold the first night, and that the total money taken in was $20,000. He said that on the second night, they sold 250 reserved-seat tickets and 1000 general-admission tickets, and that the total money taken in that night was $23,000.

To satisfy the curiosity of the Pebbles' manager about the producer's pricing policy, find out the cost of reserved-seat tickets and the cost of general-admission tickets. Include a pair of equations that will describe the problem.

# Homework 25    "Producing Programming Problems" Write-up

Your group should now have completed development of its own linear programming problem and prepared its presentation for the class. Your homework tonight is to complete your own write-up for the assignment. As with your group's presentation, this should include three things.

- A statement of the problem
- The solution
- A proof that the solution is the best possible

# Homework 26                    Continued Portfolio
                                          Selection

In *Homework 23: Beginning Portfolio Selection,* you looked at one of the major themes of this unit—linear programming problems. In this assignment, you will look at another major theme—solving systems of linear equations with two variables.

1. Summarize what you learned about solving such systems, both in *Get the Point* and in work since that activity.

2. Choose two examples of problem situations that you could solve using a system of linear equations. Your examples can be from this unit or from an earlier unit. For each of the examples you give, explain how the algebraic representation of the problem using linear equations would help you solve the problem.

# Homework 27                                           *Cookies* Portfolio

Now that *Cookies* is completed, it is time to put together your portfolio for the unit.
Compiling this portfolio has three parts.

- Writing a cover letter in which you summarize the unit
- Choosing papers to include from your work in this unit
- Discussing your personal growth during the unit

## *Cover Letter for "Cookies"*

Look back over *Cookies* and describe the central problem of the unit and the main
mathematical ideas. This description should give an overview of how the key ideas
were developed and how they were used to solve the central problem.

*Continued on next page*

# Selecting Papers from "Cookies"

Your portfolio for *Cookies* should contain

- *Homework 23: Beginning Portfolio Selection*
  Include the activities from the unit that you selected in *Homework 23: Beginning Portfolio Selection,* along with your written work about those activities that was part of the homework.

- *Homework 26: Continued Portfolio Selection*

- A Problem of the Week
  Select one of the three POWs you completed during this unit (*A Hat of a Different Color* or *Kick It!* or *Shuttling Around*).

- *Homework 17: A Reflection on Money*

- *Get the Point*

- *"How Many of Each Kind?" Revisited*

- *Producing Programming Problems*
  Include the statement and solution of the problem your group invented.

# Personal Growth

Your cover letter for *Cookies* describes how the unit develops. As part of your portfolio, write about your own personal development during this unit. You may want to address this question:

> *How do you think you have grown so far in the area of making presentations?*

You should include here any other thoughts you might like to share with a reader of your portfolio.

# Appendix

# *Supplemental Problems*

Much of this unit concerns the use of graphs to understand real-life problems. The supplemental problems for the unit continue the work with graphs and problem solving, and also follow up on some of the POWs. Here are some examples.

- *Find My Region* and *Algebra Pictures* are playful ways in which to explore graphs of equations and inequalities.

- *Rap Is Hot!* is another variation on the situation from *Homework 12: Rock 'n' Rap.*

- *Who Am I?* and *Kick It Harder!* continue the ideas from *POW 11: A Hat of a Different Color* and *POW 12: Kick It!*

# *Find My Region*

This activity is a game for two people, so the first step is to find a partner.

## *Setting Up a Feasible Region*

Each of you needs to define a feasible region using an inequality.

You should both start with the square region in the first quadrant bounded by the inequalities $x \geq 0$, $x \leq 10$, $y \geq 0$, and $y \leq 10$. This is the shaded area shown in the graph at the right.

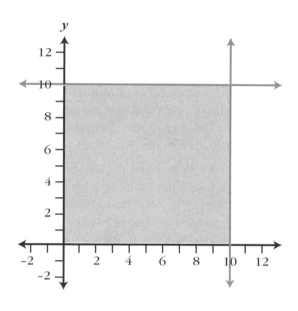

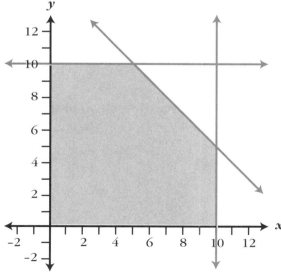

Then each of you needs to choose an inequality to restrict the region further. For example, if you choose the inequality $x + y \leq 15$, your new region will be the shaded area in the graph at the left.

You should sketch your region on a sheet of graph paper.

*Continued on next page*

# Guessing Each Other's Inequalities

Do not tell your partner what your inequality is or show him or her the region you have created. The goal of the game is to figure out what inequality the other player has used.

Sit back to back with your partner so that neither of you can see the other's region. You will each need a blank piece of graph paper to keep track of information you gather about your partner's region.

Here are the rules.

- Take turns guessing a point in your partner's region. For example, you might say, "I guess $(3, 6)$." (Because you both start with the inequalities $x \geq 0$, $x \leq 10$, $y \geq 0$, and $y \leq 10$, you should only guess points within the square region these inequalities define.)

- Each time one of you guesses a point, the other player will say "inside," "outside," or "boundary," depending on whether the point guessed is inside the region, outside the region, or on one of the boundary lines of the region. For instance, for the region sketched above, $(9, 6)$ would be a boundary point, $(5, 7)$ would be an inside point, and $(8, 9)$ would be an outside point.

- When it is your turn to guess a point, you can choose instead to guess your partner's inequality. For example, you might say, "I guess that $x + y \leq 15$ is your inequality." Your partner must answer truthfully whether your guess is equivalent to his or her inequality. If it is not, it becomes your partner's turn. You do not get a chance to guess a point.

- The winner of the game is the first player to correctly guess the other player's inequality.

*Important:* When the other player guesses an inequality, you must check whether it is equivalent to the inequality you used. For example, if the other player guesses "$x + y \leq 15$" and you used "$2x + 2y \leq 30$," you must say that this is a correct guess.

# Advanced Version

You can make this game more challenging by having each player choose two or three inequalities (in addition to the four inequalities defining the square).

# *Algebra Pictures*

You have been using pictures to help with problems in algebra. In this activity, you will use algebraic inequalities to make artistic pictures (or at least pictures someone might be interested in looking at). But one important change is that these inequalities are no longer all linear.

For each of Questions 1 through 3, make a picture by showing the solution set for the given system of inequalities.

1. $y \leq x + 8$

   $y \leq 16 - x$

   $y \geq (x - 4)^2$

2. $y \leq 2x + 4$

   $y \leq 28 - 2x$

   $y \geq 2$

   $y \leq 10$

3. $y \leq \sqrt{16 - x^2}$

   $y \geq -\sqrt{16 - x^2}$

   $y \leq \frac{1}{2}x^2$

4. Create an interesting picture using systems of inequalities. Your picture can be made up of several parts, such that each individual part is the solution set for a system of inequalities.

# *Who Am I?*

*At a college class reunion from dear old Big State U,*

*I met fifteen classmates, counting men and women, too.*

*More than half were doctors, and the rest all practiced law.*

*Of the doctors, more were females, and that I clearly saw.*

*Even more than female doctors were females doing law,*

*And these statements all would still be true including me, I saw.*

*If my friend (a noted lawyer) had a wife and kids at home,*

*Can you draw any conclusions about me from this poem?*

Prove whether you can draw any conclusions about the author from this poem. Show that the line about the friend of the author is needed.

Adapted from *MATHEMATICS: Problem Solving Through Recreational Mathematics,* by B. Averbach and O. Chein, Freeman, 1980, p. 93.

# *More Cereal Variations*

You've seen some problems involving the Hernandez twins and their breakfast habits. In this activity, your task is to make up some variations of your own.

1. Make up a variation on the situation in which the twins would choose to eat just Sweetums.

2. Make up a variation on the situation for which there would be no solution.

# *Rap Is Hot!*

Well, the distributor for Hits on a Shoestring has changed her mind about rap. That is, she has come to believe that rap is more popular in her territory than rock. She now tells the company that it can make up to twice as many rap CDs as rock CDs.

The rest of the facts are the same as in the original *Homework 12: Rock 'n' Rap* problem. Here is a summary of those constraints.

- It costs an average of $15,000 to produce a rock CD and an average of $12,000 to produce a rap CD.

- It takes about 18 hours to produce a rock CD, and about 25 hours to produce a rap CD.

- Hits on a Shoestring must use at least 175 hours of production time.

- Hits on a Shoestring can spend up to $150,000 on production costs next month.

- Each rock CD makes a profit of $20,000 and each rap CD makes a profit of $30,000.

Find out how many CDs of each type Hits on a Shoestring should make next month to maximize its profits, and justify your reasoning. (*Remember:* The company can plan to make a fraction of a CD next month and finish it the month after.)

# How Low Can You Get?

Anya and Jesse were working together on a problem, trying to find which point in a certain feasible region minimized various linear expressions involving $x$ and $y$. Here are the constraints they had.

$y \geq 3$

$y \leq 3x$

$y \leq \frac{1}{2}x + 5$

$y \leq 23 - 4x$

The feasible region for this set of constraints is shown in the graph below.

1. Invent two linear expressions in terms of $x$ and $y$ that would have different points in the region as their minimum. Give a convincing argument that the two points really are the minimums for each expression in that region.

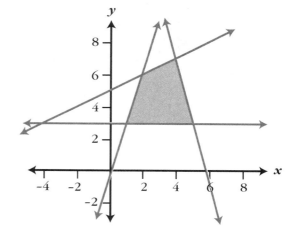

2. Invent two different linear expressions that would have the same point in the region as their minimum. Give a convincing argument that the point really is the minimum of each expression in that region.

3. Anya claims that one can pick any point in the region and find a linear expression that has its minimum for the region at that point. Jesse thinks that only certain points can be used as the minimum for a linear expression.

   Who do you think is right? If you agree with Anya, show that any point can be used as a minimum. If you agree with Jesse, explain which points can be used as a minimum and why others cannot.

# *Kick It Harder!*

In *POW 12: Kick It!* you probably found that for some scoring systems, there is a highest impossible score, and for others, there is not. In this activity, your task is to explore this issue further.

1. Find a rule that tells you which scoring systems have no highest impossible score.

2. Find a rule that tells you what the highest impossible score is when there is one.

3. Write a proof of why there is no highest impossible score for those systems for which there is none.

# Shuttling Variations

In *POW 13: Shuttling Around,* you are asked to examine a family of puzzles involving the interchange of two sets of markers. In the POW, the number of plain markers is equal to the number of shaded markers, and there is exactly one empty square in the middle. In this activity, you will consider variations on this family of puzzles.

What if the number of markers of each type can be different? For instance, you might consider the following initial situation, with three plain markers and four shaded markers.

Also, what if there is more than one empty square? For instance, you might consider the following initial situation, with two plain markers, three shaded markers, and two empty squares.

As in the POW, the task is to move the markers so that the shaded markers end up at the left and the plain markers end up at the right, according to these rules.

- The plain markers move only to the right and the shaded markers move only to the left.
- A marker can move to an adjacent open square.
- A marker can jump over *one* marker of the other type into an open square.
- No other types of moves are permitted.

As with the POW itself, you should consider these aspects of the puzzles.

- Determine whether all such puzzles have solutions.
- Look for a rule that describes the minimum number of moves in terms of the number of markers of each type and the number of empty squares.
- Prove your results.

# And Then There Were Three

You have done a lot of work in *Cookies* with two linear equations in two unknowns. In this activity, you will extend the ideas and techniques you learned to systems of three linear equations in three unknowns.

1. Here is a problem to solve with three equations in three unknowns.

   > I have some dimes, nickels, and quarters. There are 18 coins in all. The total number of dimes and nickels is equal to the number of quarters. The value of my coins is $2.85.

   > How many coins of each kind do I have?

   a. Define your variables carefully.

   b. Write three linear equations that express the facts in the situation.

   c. Solve your system of equations.

2. Make up problems for two other situations that can be solved using three variables and three linear equations.

3. Make up two more systems of three linear equations in three unknowns and try to solve them. (You do not need to make up problem situations for these systems.)

4. Write down general directions for finding the common solution for a system of three linear equations in three unknowns.

# An Age-Old Algebra Problem

Consider this problem:

> Bob, Maria, and Shoshana all have birthdays on the same day. Bob's present age is two years less than the sum of Shoshana's and Maria's present ages.
>
> In five years, Bob will be twice as old as Maria will be then. Two years ago, Maria was half as old as Shoshana was.
>
> How old is each of them?

1. Define appropriate variables and set up a system of linear equations for this problem. Be especially careful in this problem about how you define your variables.

2. Once you have written the equations, solve them using algebra.

Adapted from *MATHEMATICS: Problem Solving Through Recreational Mathematics,* by B. Averbach and O. Chein, Freeman, 1980, p. 72.

# All About Alice

# Days 1-2

## Who's Alice?

Once upon a time, a man wrote a story about the adventures of an imaginary girl named Alice, who traveled to a place called Wonderland. This story became the best-selling children's book of all time in England. The author used the pen name Lewis Carroll, and he wrote books about mathematical logic in addition to his fiction about Alice. In this unit, one of Alice's adventures forms the basis for you to explore some ideas about numerical operations, graphs, and algebraic formulas.

*Discussing homework, Mara Campbell, Nick Cabrul, and Amy McDonald compare the graphs they made of Alice's height changes.*

# Alice in Wonderland

In 1865, a book was published that was to become the most popular children's book in England—*Alice's Adventures in Wonderland.* The author used the pen name Lewis Carroll, but his real name was Charles L. Dodgson (1832–1898). He also wrote *Through the Looking Glass, and What Alice Found There,* a sequel to his original Alice story.

*Continued on next page*

# The Situation of Alice

Here is an excerpt from *Alice's Adventures in Wonderland*. Read the excerpt, and then answer the questions below.

> [Alice] found a little bottle . . . , and tied round the neck of the bottle was a paper label with the words "DRINK ME" beautifully printed in large letters. . . .
>
> So Alice ventured to taste it, and, finding it very nice . . . , she very soon finished it off.
>
> "What a curious feeling!" said Alice. "I must be shutting up like a telescope!"
>
> And so it was indeed: she was now only ten inches high. . . .
>
> Soon her eye fell on a little glass box that was lying under the table: she opened it, and found in it a very small cake, on which the words "EAT ME" were beautifully marked in currants. . . .
>
> She ate a little bit . . . and very soon finished off the cake.
>
> "Curiouser and curiouser!" cried Alice. . . . "Now I'm opening out like the largest telescope that ever was! Goodbye, feet!" (for when she looked down at her feet, they seemed to be almost out of sight, they were getting so far off). . . .
>
> Just at this moment her head struck against the roof of the hall.

# The Questions

In Lewis Carroll's story, whenever Alice drinks the beverage from the bottle, she gets smaller, and when she eats the cake, she gets bigger. But Carroll doesn't say *how much bigger* or *how much smaller*, or even *how tall Alice was to start with*.

Assume that for every ounce of the cake Alice eats, her height doubles, and for every ounce of the beverage she drinks, her height is cut in half. Answer these questions based on that assumption.

1. What happens to Alice's height if she eats 2 ounces of cake? What if she eats 5 ounces?

2. Find a rule for what happens to Alice's height when she eats $C$ ounces of cake. Explain your rule.

3. What happens to Alice's height if she drinks 4 ounces of beverage? What if she drinks 6 ounces?

4. Find a rule for what happens to Alice's height when she drinks $B$ ounces of beverage. Explain your rule.

# POW 14 *More from Lewis Carroll*

Lewis Carroll was a mathematician as well as a novelist. One of his special mathematical interests was **logic,** which might be described as the mathematical science of formal reasoning. Logic analyzes how to draw legitimate, or *valid*, conclusions from true statements. This process of drawing conclusions is also called **deduction.**

In one of Lewis Carroll's books, he gave problems involving groups of statements. The reader was supposed to figure out what, if anything, could be deduced from the statements, that is, what new conclusions could be drawn. All of the groups of statements in this POW are taken directly from Lewis Carroll's work.

Here are two examples.

## *Example 1*

a. John is in the house.

b. Everyone in the house is ill.

If you know that statements a and b are both true, then you can deduce that John must be ill. So "John is ill" is a valid conclusion.

*Continued on next page*

# Example 2

    a. Some geraniums are red.

    b. All these flowers are red.

In this case, knowing that statements a and b are both true does not tell you whether any or all of "these flowers" are geraniums. They might be other kinds of red flowers. So there isn't anything new that you can definitely deduce from the two statements in Example 2.

# Part I: Finding Conclusions

Examine each of the sets of statements given here. Decide what, *if anything,* you could deduce if you knew that the given statements were true. (There may be more than one conclusion possible. Give as many conclusions as you can.)

Explain in each case why you think your conclusions are correct or why you think no new conclusions can be deduced. Diagrams or pictures might be helpful both in analyzing the statements and in explaining your reasoning.

1. a. No medicine is nice.

    b. Senna is a medicine.

2. a. All shillings are round.

    b. These coins are round.

3. a. Some pigs are wild.

    b. All pigs are fat.

4. a. Prejudiced persons are untrustworthy.

    b. Some unprejudiced persons are disliked.

5. a. Babies are illogical.

    b. Nobody who is despised can manage a crocodile.

    c. Illogical persons are despised.

6. a. No birds, except ostriches, are 9 feet high.

    b. There are no birds in this aviary that belong to anyone but me.

    c. No ostrich lives on mince pies.

    d. I have no birds less than 9 feet high.

*Continued on next page*

## Part II: Creating Examples

Make up two sets of statements similar to those in Part I. One of your sets should have a valid conclusion, and the other should not.

## Write-up

1. *Process*

2. *Results:* Give your conclusions (with explanations) for each set of statements in Part I. Give your sets of statements for Part II and explain why they do or do not have valid conclusions.

3. *Evaluation:* What does this POW have to do with mathematics?

4. *Self-assessment*

Examples 1 and 2 and the statements in Part I are taken from *Symbolic Logic and The Game of Logic,* by Lewis Carroll, Dover Publications, Inc., New York and Berkeley Enterprises, 1958.

# Homework 1                    Graphing Alice

In the activity *Alice in Wonderland*, you looked at what happened to Alice's height in various situations. This assignment involves looking at that information in an organized way.

Choose a suitable scale for each of the graphs you create. (*Note:* The scales of the two axes do not need to be the same.)

1. Alice's height changes when she eats the cake. Assume as before that her height doubles for each ounce she eats.

    a. Find out what Alice's height is multiplied by when she eats 1, 2, 3, 4, 5, or 6 ounces of cake.

    b. Make a graph of this information.

2. Alice's height also changes when she drinks the beverage. Assume as before that her height is halved for each ounce she drinks.

    a. Find out what Alice's height is multiplied by when she drinks 1, 2, 3, 4, 5, or 6 ounces of beverage.

    b. Make a graph of this information.

3. Suppose Alice found a different kind of cake, one that tripled her height for every ounce consumed. Do Questions 1a and 1b for this different cake.

4. Suppose Alice found a different kind of beverage, one that cut her height to one-third of its measure for every ounce consumed. Do Questions 2a and 2b for this different beverage.

5. Compare and contrast the graphs from Questions 1 through 4. In general, what do you think is true of graphs such as these?

# Homework 2                          A Wonderland Lost

The Amazon rain forest is gradually being destroyed by pollution and agricultural and industrial development. Suppose, for simplicity, that each year, 10 percent of the remaining forest is destroyed. Assume for this assignment that the present area of the Amazon rain forest is 1,200,000 square miles.

1. a. What will the area of the forest be after one year of this destruction process?

   b. What will the area of the forest be after two years of the destruction process?

2. Make a graph showing your results from Question 1 and continuing through five years of the destruction process. Include the present situation as a point on your graph.

3. Find a rule for how much rain forest will remain after $X$ years. That is, express the area of the rain forest as a function of $X$.

4. Explain how this situation and its graph relate to Alice and her situation.

# Days 3-8

# *Extending Exponentiation*

The fundamental idea in Alice's adventure with the strange cake and beverage is that she grows and shrinks exponentially. Of course, the definition of exponentiation as repeated multiplication requires that the exponent be a positive whole number. But what if the exponent is zero? Or negative?

In the next portion of the unit, you'll use Alice's situation to gain insight into how the operation of exponentiation can be extended to allow these new types of exponents.

*Erica Lanetot, Molly Jansen, Danielle Crisler, and Jillian Clark compare the results they got using the additive law of exponents.*

# *Here Goes Nothing*

For these problems, the cake Alice eats is base 2 cake, as in the original problem.

1. What would happen to Alice's height if she ate 0 ounces of cake? Specifically, what would her height be multiplied by?

2. In *Homework 1: Graphing Alice,* you made a graph of what Alice's height is multiplied by as a function of how much base 2 cake she eats. Examine that graph and explain whether your answer to Question 1 makes sense for that graph.

3. In *Alice in Wonderland,* you developed the general rule that eating $C$ ounces of base 2 cake multiplies Alice's height by $2^C$. According to this rule, what should Alice's height be multiplied by if she eats 0 ounces of this cake?

4. What does all this seem to point to as the value of $2^0$?

# Homework 3        A New Kind of Cake

## Part I: Base 5 Cake

Alice has discovered base 5 cake; that is, each ounce she eats of this cake multiplies her height by 5.

1.  Figure out what Alice's height would be multiplied by if she ate 1 ounce, 2 ounces, 3 ounces, or 4 ounces of this cake.

2.  Make a graph of your results from Question 1, using $x$ for the number of ounces Alice eats and $y$ for the number her height is multiplied by.

3.  a.  What would Alice's height be multiplied by if she ate 0 ounces of cake?

    b.  Consider your answer to Question 3a in connection with your graph from Question 2. Does your answer to Question 3a seem to fit as a likely value for $y$ when $x$ is 0?

    c.  Explain what Questions 3a and 3b have to do with defining $5^0$.

4.  Use a pattern approach to explain why it makes sense to say $5^0 = 1$.

## Part II: Base or Exponent?

5.  Graph the equations $y = 2^x$ and $y = x^2$ on the same set of axes. As $x$ gets big, which graph has the larger $y$-value? (Be sure to plot enough points to get a sense of the growth of each function.)

# Piece After Piece

Alice does not always eat her cake in one sitting. At times, she eats some cake, follows a rabbit for a while, and then comes back and eats some more cake. In this activity, you investigate what happens to Alice's height when she eats piece after piece of cake. (Assume that she has the original base 2 cake.)

1. Suppose Alice eats a 3-ounce piece of cake, takes a break, and then later eats a 5-ounce piece of cake.

   a. What happens to her height?

   b. Is the result the same as if she had eaten a single 8-ounce piece of the cake? Explain.

2. Make up two more pairs of questions like those in Question 1. Answer them and explain your reasoning.

3. Now answer Questions 1 and 2 as if they were about the beverage instead of about the cake.

# Homework 4 When Is Nothing Something?

Clarabell says:

> "The number 0 stands for nothing. So $3^0$ means no 3's. No 3's is zero, so $3^0$ equals 0."

Bellaclar says:

> "The number 0 stands for nothing. So $3^0$ is the same as a 3 with no exponent, and that's just 3. Therefore, $3^0$ equals 3."

1. Explain to Clarabell and Bellaclar why $3^0$ is not equal to 0 or to 3.

2. Here are two situations in which the number 0 is *not* nothing.

   • Example 1: The 0 in 20 is not nothing, because otherwise there would be no difference between 2 and 20.

   • Example 2: A temperature of 0 degrees is not the same as there not being any temperature (whatever that means).

   Make up two more situations in which 0 is *not* nothing.

3. Mathematicians have decided that it makes sense to define $3^0$ (and expressions like it) as equal to 1. They saw that this definition fits well with other principles about exponents, and so it makes working with exponents logical.

   When people agree to use a word or symbol in a particular way, such an agreement is called a **convention.** Think back on other mathematical topics you have studied, and write about another situation in which you think a *convention* is involved.

# *Many Meals for Alice*

Alice has decided that she will be healthier if she eats fewer sweets. Therefore, she will eat a fixed number of ounces of cake each time she sits down for a meal.

Your task is to find out what would happen to her height after different numbers of meals with a given amount of cake. (Alice is eating base 2 cake in these problems.)

1. Suppose Alice decides that she will eat 3 ounces of cake at each meal. What will her height be multiplied by after two meals? After three meals? After four meals? After *M* meals?

2. Experiment with different amounts of cake at each meal and different numbers of meals. Use your examples to develop an expression for what her height will be multiplied by after *M* meals with *D* ounces of cake at each meal.

3. Would eating 4 ounces of cake at each of six meals be the same as eating 6 ounces of cake at each of four meals? Why or why not?

4. Make up another example like Question 3. Your example should compare two situations in which you switch the number of ounces with the number of meals.

5. Write a general law of exponents that expresses your observations from Questions 3 and 4.

# Homework 5        In Search of the Law

You have seen that if you multiply two exponential expressions with the same base, such as $2^3$ and $2^5$, the product is an expression such as $2^8$, where the base is the same as before and the exponent is the sum of the exponents from the factors.

This principle is called the **additive law of exponents** and can be expressed by the general equation

$$A^X \cdot A^Y = A^{X + Y}$$

Actually, there are many laws that relate to exponents. In this assignment, you'll investigate other possible laws.

1. Suppose the exponential expressions being multiplied have different bases but the same exponent. That is, consider products of the form $A^X \cdot B^X$. Look for a general law for simplifying such products. As needed, examine specific cases, such as $3^7 \cdot 5^7$, using the definition of exponentiation to write this expression as a product of 3's and 5's.

2. Suppose the two factors have the same base *and* the same exponent. Do you apply the additive law of exponents or do you use your answer to Question 1? Look at specific cases to investigate what to do with expressions of the form $A^X \cdot A^X$.

3. A common mistake people make when working with exponents is to multiply the base by the exponent instead of raising the base to the power of the exponent. For instance, they would mistakenly say that $2^3$ is 6 (because $2 \cdot 3 = 6$) when the correct answer is actually 8 (because $2 \cdot 2 \cdot 2 = 8$).

   Are there any pairs of numbers for which this mistake in thinking actually gives the correct answer? In other words, are there any solutions to the equation $A^X = A \cdot X$? If so, what are they?

# *Having Your Cake and Drinking Too*

You have found that the additive law of exponents gives an easy way to calculate what happens to Alice when she eats several pieces of cake. For example, the equation below shows how to combine the effect of a 17-ounce piece of base 2 cake with that of a 5-ounce piece of that cake.

$$2^{17} \cdot 2^5 = 2^{17 + 5} = 2^{22}$$

You also found out how to combine several servings of the beverage. For example, here is an equation that shows how to combine the effect of a 4-ounce serving of base 2 beverage with that of a 9-ounce serving of that beverage.

$$\left(\frac{1}{2}\right)^4 \cdot \left(\frac{1}{2}\right)^9 = \left(\frac{1}{2}\right)^{4 + 9} = \left(\frac{1}{2}\right)^{13}$$

*Continued on next page*

In this activity, your goal is to figure out how to determine the effect on Alice of combining base 2 cake and base 2 beverage.

1. What is Alice's height multiplied by if she consumes the same number of ounces of cake and beverage? Write an equation using exponential expressions that expresses your answer.

2. Write at least five ways to combine eating cake with drinking beverage that will result in Alice being 8 times her original height. That is, find combinations of amounts of cake and of beverage for which her original height will be multiplied by $2^3$.

3. a. Find several combinations of amounts of cake and beverage that will result in Alice being 32 times her original height.

   b. Find several combinations of amounts of cake and beverage that will result in Alice being 4 times her original height.

4. Look for a pattern in your answers to Questions 2 and 3. Write a general expression for the amount Alice's height is multiplied by if she eats $C$ ounces of cake and drinks $B$ ounces of beverage.

5. What happens to your rule in Question 4 if $B$ is more than $C$?

# Homework 6    Rallods in Rednow Land

The ruler of Rednow Land had a very wise advisor who had saved the country in various ways (such as finding counterfeiters of gold coins). The ruler wanted to reward this wise person.

The ruler loved to play chess and so came up with two choices of rewards for the wise advisor.

- • Choice A: A billion rallods (a rallod is the coin of Rednow Land)

- • Choice B: The amount of money obtained by putting 1 rallod on one square of the chessboard, 2 rallods on the next, 4 on the next, 8 on the next, and so on until all 64 squares were filled

1. What does your intuition tell you about which would be the better choice?

2. Now make a decision based on the results of computation. Explain your decision.

3. The standard chessboard has 64 squares. How many squares would be needed to make Choice B as close as possible to Choice A? Explain your reasoning.

*Historical note:* This assignment is an adaptation of a problem that can be traced to Persia in about the seventh century and that may have originated even earlier in India.

# Homework 7     Continuing the Pattern

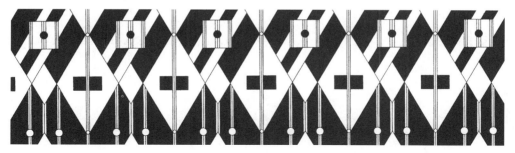

*Closeup of an Art Deco design from "Original Art Deco Allover Patterns," copyright*
*© 1989 by Dover Publications, Inc.*

You've seen several different ways to explain the definition of $2^0$ as 1. In this
assignment, you'll adapt one of those methods to think about defining exponential
expressions using negative exponents.

1. Begin by examining the powers of 2 shown here.

$$2^5 = 32$$
$$2^4 = 16$$
$$2^3 = 8$$
$$2^2 = 4$$
$$2^1 = 2$$
$$2^0 = 1$$
$$2^{-1} = ?$$
$$2^{-2} = ?$$
$$2^{-3} = ?$$
$$2^{-4} = ?$$

   a. Describe the pattern of numerical values shown on the right sides of these
equations for powers of 2 with positive and zero exponents.

   b. Explain how you would use this pattern to find the missing values for
powers of 2 with negative exponents. Express your numerical results *as
fractions* (not as decimals).

*Continued on next page*

2. a. Make similar lists for powers of 3 and powers of 5 using positive and zero exponents, and extend these lists to negative exponents.

   b. Describe how your results for these lists compare with your results on Question 1.

3. If the base is $\frac{1}{2}$, the list of powers for positive exponents looks like this.

$$\left(\tfrac{1}{2}\right)^5 = \tfrac{1}{32}$$

$$\left(\tfrac{1}{2}\right)^4 = \tfrac{1}{16}$$

$$\left(\tfrac{1}{2}\right)^3 = \tfrac{1}{8}$$

$$\left(\tfrac{1}{2}\right)^2 = \tfrac{1}{4}$$

$$\left(\tfrac{1}{2}\right)^1 = \tfrac{1}{2}$$

$$\left(\tfrac{1}{2}\right)^0 = 1$$

Extend this list to negative exponents, and compare your results with your results on Questions 1 and 2.

# Homework 8                    Negative Reflections

When you first learned about exponents, their use was defined in terms of repeated multiplication. For example, you defined $2^5$ as $2 \cdot 2 \cdot 2 \cdot 2 \cdot 2$.

With that repeated-multiplication definition, the exponent had to be a positive whole number. Now you have seen a way to make new definitions that allow zero and negative integers to be exponents.

1. Write a clear explanation summarizing what you have learned about defining expressions that have zero or a negative integer as an exponent. Then explain, using examples, why these definitions make sense. Give as many different reasons as you can, and indicate which explanation makes the most sense to you.

2. Show your explanation to an adult, and ask that person whether it made sense to him or her. Then write about the person's reaction.

# Days 9-12

# *"Curiouser and Curiouser!"*

Alice thought it was "a curious feeling" when she began shrinking as a result of drinking her special beverage. You might have thought it was rather curious when you learned that zero and negative integers could be used as exponents.

"Curiouser and curiouser!" That's what Alice said when she learned about the special cake. Your adventures with exponents also get curiouser, as you go from integers to fractions in the next part of the unit.

*In "All Roads Lead to Rome," Adam Davenport, Dean Mertes, Tiffany Tomlin, and Joe Stebbins review the extension of exponentiation beyond positive integral exponents.*

# A Half Ounce of Cake

One day, as Alice was wandering through Wonderland, she stumbled across a silver plate with a small piece of cake on it. Alice picked up the cake and could tell by the size and feel of it that it weighed exactly half an ounce. She also could tell from the aroma and texture that this was base 2 cake.

1. We all know that eating an ounce of this cake will double Alice's height. But what will eating half an ounce multiply her height by? (*Hint:* Keep in mind that eating half an ounce of cake and then eating another half ounce should have the same effect as eating one ounce of cake.)

2. Investigate what Alice's height is multiplied by if she eats other fractional pieces of cake, such as a third of an ounce or a fifth of an ounce.

# Homework 9 It's in the Graph

What's $2^{1/2}$?

Maybe you know and maybe you don't. If you don't know, a graph could help you find out. If you do know, a graph will give you another way of thinking about that number.

In *Homework 1: Graphing Alice,* you made a graph showing what Alice's height is multiplied by if she eats various amounts of cake. As you have seen, that graph showed points that fit the equation $y = 2^x$.

In that graph, you considered only positive integers for $x$. You now know how to interpret the expression $2^x$ when $x$ is any integer.

1. a. Make an In-Out table for the equation $y = 2^x$ using the values $-2, -1, 0, 1,$ and $2$ for $x$. Then plot the points from your table and connect them with a smooth curve.

   b. Use your graph from part a to estimate the value of $2^{1/2}$.

*Continued on next page*

The curve you drew in Question 1b went through the specific points (0, 1) and (1, 2). Question 2 deals with the graph of the equation of the *line* through these two points.

2. a. Draw the graph of the equation $y = x + 1$ on the same axes you used for Question 1.

    b. Compare the two graphs. What does this comparison tell you about the value of $2^{1/2}$?

3. Parts a, b, and c below are similar to the process you used in Question 1.

    a. Make an In-Out table for the equation $y = 3^x$ using the values $-2, -1, 0, 1,$ and 2 for $x$. Then plot the points, connect them with a smooth curve, and use your graph to estimate the value of $3^{1/2}$.

    b. Use a similar process to estimate the value of $9^{1/2}$ by making a table for the equation $y = 9^x$, plotting and connecting the points, and estimating.

    c. Use a similar process to estimate the value of $\left(\frac{1}{2}\right)^{1/2}$, using the equation $y = \left(\frac{1}{2}\right)^x$.

# POW 15     *A Logical Collection*

This POW is somewhat different from most in that it contains three separate problems. What these problems have in common is that they involve logical reasoning to figure out who is telling the truth.

Your write-up for each problem should explain how you solved it and how you can prove that your answers are correct.

## Part I: The Missing Mascot

The mascot of Goldenrod High is a stuffed ostrich, and it sits outside the main office. Just before the big game, the ostrich disappeared, and three students from arch-rival Greenview High are being questioned.

Each of the suspected students has made some statements.

Adams said:

- "I didn't do it."
- "Benitez was hanging out near Goldenrod that day."

Benitez said:

- "I didn't do it."
- "I've never been inside Goldenrod."
- "Besides, I was out of town all that week."

Clark said:

- "I didn't do it."
- "I saw Adams and Benitez near Goldenrod that day."
- "One of them did it."

Assume that two of these students are innocent and are telling the truth, but that the third student is guilty and may be lying. Who did it? Prove your answer.

*Continued on next page*

---

## *Part II: What Did He Say?*

You have found a strange place where some people always tell the truth and the rest always lie. Unfortunately, there's no way to tell from looking at them which is which.

You find yourself sitting with three of these people and decide to try to determine who belongs to which category. For simplicity, we'll refer to the three people as A, B, and C.

Here's the conversation.

> You say to A: "Are you a truth-teller or a liar?"
>
> A answers your question, but a squawking bird prevents you from hearing the answer.
>
> B says: "A says he's a truth-teller."
>
> C says: "B is lying."

What can you figure out from this conversation? What can't you figure out? Prove your answers.

## *Part III: The Turner Triplets*

The Turner triplets have a policy that whenever anyone asks them a question, two of them tell the truth and the other one lies. You have just asked them all which one was born first.

Here are their answers.

> Virna: "I am not the oldest."
>
> Werner: "Virna was born first."
>
> Myrna: "Werner is the oldest."

Who was born first? Prove your answer.

# Homework 10    Stranger Pieces of Cake

In the activity *A Half Ounce of Cake,* you saw how to use Alice's situation to make sense of certain fractional exponents, such as $\frac{1}{2}$ and $\frac{1}{3}$, where the numerator of the fraction is 1. (These are called **unit fractions.**) But what about fractional exponents in general?

In this assignment, you will investigate the effect of fractional pieces of cake where the fraction's numerator is not 1.

1. Start with a piece of base 2 cake that weighs $\frac{3}{5}$ of an ounce. What effect should eating this have on Alice's height? Explain your answer.

2. Use your work on Question 1 to give a general way of defining $2^{p/q}$ for any fraction $\frac{p}{q}$. Explain your ideas.

# Homework 11      Confusion Reigns

The students at Bayside High School are learning some fancy stuff about exponents, and there seems to be some confusion. They are trying to come up with some other generalizations besides the additive law of exponents.

1. Bill says,

   "$3^4 + 3^5 = 3^9$, because $4 + 5 = 9$."

   Jill says,

   "$3^4 + 3^5 = 6^9$, because $4 + 5 = 9$ and $3 + 3 = 6$."

   Do either of them know what's going on? Don't simply give yes or no answers. Bill and Jill need some good explanations if they are going to understand how to work with exponents.

*Continued on next page*

2. Randy, Sandy, and Dandy each say that they have a way to multiply expressions with the same exponent.

Randy says,

"$2^3 \cdot 5^3 = 10^6$, because $2 \cdot 5 = 10$ and $3 + 3 = 6$."

Sandy says,

"$2^3 \cdot 5^3 = 10^9$, because $2 \cdot 5 = 10$ and $3 \cdot 3 = 9$."

Dandy says,

"$2^3 \cdot 5^3 = 10^3$, because $2 \cdot 5 = 10$ and the exponent doesn't change."

Again, do any of them know what's going on? And, again, don't simply give yes or no answers. Try to come up with a particular rule for multiplying expressions with exponents when the exponents are the same, and give good explanations for your answers.

3. Fran, Jan, and Stan want to raise exponential expressions to powers.

Fran says,

"$\left(7^2\right)^3 = 7^5$, because $2 + 3 = 5$."

Jan says,

"$\left(7^2\right)^3 = 7^6$, because $2 \cdot 3 = 6$."

Stan says,

"$\left(7^2\right)^3 = 7^8$, because $2^3 = 8$."

Once again, who's right and who's wrong? And, once again, don't simply give yes or no answers. Try to come up with a rule for raising exponential expressions to powers, and give a good explanation for your answer.

# All Roads Lead to Rome

The basic definition for exponential expressions is given in terms of repeated multiplication. For example, $3^5$ means "multiply five 3's together." This gives $3 \cdot 3 \cdot 3 \cdot 3 \cdot 3$, which is 243. Thus, $3^5 = 243$.

In this definition, the **exponent** tells you how many of the **bases** to multiply together. This definition makes sense when the exponent is a positive integer. But you can't interpret zero, negative, or fractional exponents in terms of how many of the bases to multiply. For example, it doesn't make sense to say "multiply negative six 3's together."

*Continued on next page*

In this unit, you have seen other ways to make sense of exponents that are not positive integers.

- You can use the Alice story.

- You can extend a numerical pattern that starts with positive integer exponents.

- You can see what definition will be consistent with the additive law of exponents, which says $A^B \cdot A^C = A^{B+C}$.

- You can make a graph of the equation $y = A^x$ using positive integer values for $x$ and use the graph to estimate $y$ for other values of $x$.

Fortunately, these different approaches lead to the same conclusions. The problems here give you a chance to show how the different methods work.

1. Suppose Alice has base 5 cake and beverage. That is, for each ounce of cake Alice eats, her height is multiplied by 5, and for each ounce of beverage she drinks, her height is multiplied by $\frac{1}{5}$. Explain the meaning of $5^0$ using all four methods described above.

2. The four methods don't necessarily all make sense for every possible exponent. Explain the meaning of each of the exponential expressions here using as many of the four ways as make sense for the particular example. You will need to decide in each case what base of cake or beverage to use.

   a. $3^{-4}$

   b. $2^{1/2}$

   c. $7^{1/3}$

   d. $32^{2/5}$

# Homework 12    Measuring Meals for Alice

In this assignment, Alice is using her original base 2 cake and beverage. That is, 1 ounce of cake doubles Alice's height and 1 ounce of beverage halves her height.

Using a scientific calculator, find answers to the nearest tenth of an ounce or tenth of a foot for each of these questions, and explain your answers.

1. If Alice is 1 foot tall, how much cake should she eat to become 10 feet tall?

2. a. If Alice is 1 foot tall, how much cake should she eat to become 100 feet tall?

    b. Compare your result in Question 2a to your answer in Question 1, and discuss the connection between the two problems.

3. If Alice is 9 feet tall and wants to be 3 feet tall, how much beverage should she drink?

4. If Alice is 20 feet tall and she drinks 2.4 ounces of beverage, how tall will she be?

**Days 13-18**

# *Turning Exponents Around*

If you know what kind of cake and how much cake Alice is eating, then you can figure out what her height will be multiplied by. But what if you only know the *kind* of cake, and you want her to grow by a certain factor? How can you figure out *how much* cake she should eat?

In the final portion of this unit, you'll explore questions like this. You'll learn some special ways to express the answers to such questions, as well as special notation for representing very big and very small numbers.

*Mario Sandoval questions Jason Torres about the graph displayed on his calculator screen.*

# Homework 13  Sending Alice to the Moon

1. Alice has just discovered base 10 cake and is delighted with how powerful it is. One day, after nibbling on the cake, Alice realized that she was 1 mile tall. Having her head up in the sky got her thinking about space travel, and she decided it would be nice to visit the moon, which is about 239,000 miles from the earth.

   How many more ounces of base 10 cake should Alice eat so that the top of her head just touches the moon? Give your answer to the nearest hundredth of an ounce.

2. After her head reached the moon, Alice continued to eat cake until her head reached the planet Pluto, which at that time was approximately 3,670,000,000 miles from the earth. But she forgot to keep track of how many ounces of cake she had eaten to get that tall.

   How many ounces of base 10 beverage must she drink in order to return to a height of 1 mile? Give your answer to the nearest hundredth of an ounce.

# Homework 14     Alice on a Log

In this assignment, Alice is thinking about base 10 cake and beverage. If she eats 1 ounce of this kind of cake, her height will be multiplied by 10, and if she drinks 1 ounce of the beverage, her height will be multiplied by $\frac{1}{10}$.

Alice has just heard about logarithms and is all excited. For example, she found out that $\log_{10} 162$ means "the power to which I should raise 10 to get 162."

"How could anyone find that exciting?" you ask.

Well, Alice thinks it sounds much more sophisticated to ask, "What is $\log_{10} 162$?" than to ask, "How many ounces of base 10 cake should I eat in order to grow to 162 times my height?"

1. Between what two whole numbers does the value of $\log_{10} 162$ lie? Explain your answer.

2. For each of these questions, write a logarithm expression that represents the answer and then find the numerical value of the expression.

   a. How many ounces of base 10 cake should Alice eat to grow to 100 times her size?

   b. How many ounces of base 10 cake should Alice eat to grow to 10,000 times her size?

   c. How many ounces of base 10 cake should Alice eat to grow to 50 times her size?

   d. How many ounces of base 10 cake should Alice eat to grow to 2000 times her size?

   e. How many ounces of base 10 beverage should Alice drink to shrink to $\frac{1}{10}$ her size?

   f. How many ounces of base 10 beverage should Alice drink to shrink to $\frac{1}{4}$ her size?

# Taking Logs to the Axes

Once Alice found out that $\log_{10} 162$ means "the power to which I should raise 10 to get 162," she got curious about what the graph of a logarithm function would look like.

She realized that there was a different logarithm function for each base. For example, one such function would be defined by the equation $y = \log_2 x$.

Her investigations relied heavily on the fact that the equation $c = \log_a b$ means the same thing as the equation $a^c = b$. Using this relationship allowed her to work with exponential equations, and she was more comfortable with them.

1. In each of parts a and b, choose values for $x$ for which you can easily compute the value of $y$, and plot the resulting points. Choose enough points in each case to allow you to sketch the entire graph.

   a. $y = \log_2 x$

   b. $y = \log_3 x$

2. Use a graphing calculator to draw the graph of the equation $y = \log_{10} x$.

3. Compare the graph of the logarithm function using base 2 with the graphs of logarithm functions using different bases. In general, how does the graph change as the base gets larger? Why?

4. How does the graph of a logarithm function compare to the graph of the corresponding exponential function?

# Homework 15          Base 10 Alice

All the questions in this assignment refer to base 10 cake and beverage.

1. For each of the quantities shown below, find Alice's height after eating that amount of cake. In each case, assume that she starts at a height of 5 feet.

   a. 4 ounces

   b. 8 ounces

   c. 13 ounces

2. If Alice eats a whole number of ounces of cake and starts from a height of 5 feet, what do you know about the possible heights she can grow to?

3. Suppose Alice is 5 feet tall and wants to know how many ounces of cake she needs to eat to become 50,000,000,000 feet tall. (That's 50 billion feet, which is roughly 10 million miles.) What shortcut can you use in answering her question?

4. Pick three different whole numbers of ounces of beverage for Alice to drink, and find her height after consuming each amount. Assume in each case that she starts out 5 feet tall.

5. Find a simple rule for writing Alice's final height for situations like those you made up in Question 4. Your rule should deal specifically with the case of whole-number ounces of beverage.

# Homework 16

# Warming Up to Scientific Notation

Scientific notation may be a new way for you to express numbers. It often takes some practice to get used to working with scientific notation, but it's worth the effort because many ideas in mathematics and science are expressed using this special way of writing numbers.

This homework assignment gives you several ways to get accustomed to scientific notation.

1. Write each of these numbers in scientific notation.

   a. 34,200

   b. 0.0034

*Continued on next page*

2. Write each of these numbers in ordinary notation.

    a. $4.2 \cdot 10^5$

    b. $7.503 \cdot 10^{-2}$

3. Each of the next series of problems gives a product or quotient of two numbers written in scientific notation. Find the numerical value of each result *without using a calculator,* and write your final answers in scientific notation.

    a. $(3 \cdot 10^4) \cdot (2 \cdot 10^7)$

    b. $(5 \cdot 10^5) \cdot (7 \cdot 10^{-2})$

    c. $(7 \cdot 10^8) \div (2 \cdot 10^3)$

    d. $(9 \cdot 10^3) \div (3 \cdot 10^{-4})$

    e. $(6 \cdot 10^{-3}) \div (2 \cdot 10^{-8})$

    f. $(2 \cdot 10^5) \div (4 \cdot 10^3)$

4. Based on the examples in Question 3, develop some general principles for multiplying and dividing numbers written in scientific notation. Make up more examples as needed, and illustrate your rules.

5. a. Figure out and describe how to *enter* numbers in scientific notation on your own scientific calculator.

    b. Describe how your scientific calculator *displays* numbers in scientific notation.

# *Big Numbers*

Scientific notation is sometimes helpful in working with big numbers.

In some of the problems in this activity, you are given information in scientific notation. You can use what you learned in *Homework 16: Warming Up to Scientific Notation* to simplify the computations.

You will probably want to write your answers in scientific notation. But you do not necessarily need to give exact answers for these problems. Use your judgment about the amount of precision that is appropriate in each case.

1. A computer can do a computation in $5 \cdot 10^{-7}$ seconds. How many computations can the computer do in 30 seconds?

2. A leaking faucet drips a drop per second. If there are 76,000 drops of water in a gallon, how many gallons would drip in a year?

*Continued on next page*

3. Measurements show that Europe and Africa are separating from the Americas at a rate of about 1 inch per year. The continents are now about 4000 miles apart. Assuming that the rate has remained constant, how many years has it been since the continents split apart and started drifting?

4. In 1990, the gross national debt of the United States was $3,233,313,000,000. The 1990 census showed 248,709,873 U.S. citizens. About how many dollars per citizen was the national debt in 1990?

5. One atom of carbon weighs approximately $1.99 \cdot 10^{-23}$ grams. How many atoms are there in a kilogram of carbon?

6. The mass of the earth is $5.98 \cdot 10^{24}$ kg. The mass of the sun is $1.99 \cdot 10^{30}$ kg. Approximately how many earths would it take to have the same mass as the sun?

7. Light travels at a speed of approximately 186,000 miles per second. (That's *very* fast.) A **light-year** is the distance that light travels in a year. Approximately how many inches are there in a light-year?

8. For simplicity, suppose that a grain of sand is a cube that is 0.2 millimeters in each direction. About how many grains of sand packed tightly together would it take to make a beach that is 300 meters long, 25 meters wide, and 5 meters deep?

Questions 1–4 were adapted from problems in *Algebra I,* by Paul A. Foerster, Addison-Wesley Publishing Co., 1990, pp. 404–406. Data in Question 4 taken from *Information Please,* 1996 Almanac, 49th edition, Houghton-Mifflin.

# Homework 17  An Exponential Portfolio

Although you have seen in this unit that exponents don't have to be positive integers, you have also seen many general laws about exponents that are based on the definition of exponentiation as repeated multiplication.

Your task in this assignment is to list all the general laws of exponents that you have studied. Give at least one explanation for each general law. Your explanations may be based on the definition or may use some other approach, such as the "Alice" metaphor or a numerical pattern.

# "All About Alice" Portfolio

Now that *All About Alice* is completed, it is time to put together your portfolio for the unit. Compiling this portfolio has three parts:

- Writing a cover letter in which you summarize the unit

- Choosing papers to include from your work in this unit

- Discussing your personal growth during the unit

## Cover Letter for "All About Alice"

Look back over *All About Alice* and describe the central theme of the unit and the main mathematical ideas. This description should give an overview of how the key ideas were developed in this unit.

As part of the compilation of your portfolio, you will select some activities that you think were important in developing the key ideas. Your cover letter should include an explanation of why you are selecting each particular item.

## Selecting Papers from "All About Alice"

Your portfolio from *All About Alice* should contain

- *Homework 17: An Exponential Portfolio*

- *All Roads Lead to Rome*

- A homework or class activity in which you used exponents to solve the problem

*Continued on next page*

- A homework or class activity that involved graphing

- *Homework 8: Negative Reflections*

- A Problem of the Week
  Select one of the two POWs you completed during this unit (*More From Lewis Carroll* or *A Logical Collection*).

# Personal Growth

Your cover letter for *All About Alice* describes how the unit develops. As part of your portfolio, write about your personal development during this unit. You may want to address this question.

> *How do you think you have grown mathematically over your second year of IMP?*

You should include here any other thoughts about your experiences with this unit that you want to share with a reader of your portfolio.

# Appendix

# *Supplemental Problems*

Most of the supplemental problems for *All About Alice* continue the focus on exponents and related ideas. Here are some examples.

- *Inflation, Depreciation, and Alice* presents the idea of how changes in prices might involve exponential functions.

- *A Little Shakes a Lot* describes an important use of logarithms.

- *Very Big and Very Small* continues your work with scientific notation.

# *Inflation, Depreciation, and Alice*

As you probably have discovered, prices tend to go up over the years. For example, you may know that it used to cost only 10¢ to use a pay phone, but it now costs 20¢ or 25¢.

This sort of rise in costs is called **inflation.** The inflation rate for an item or service is the percentage increase in its price over time. For instance, an annual inflation rate of 5 percent means that prices go up 5 percent each year. (Of course, the rate of inflation is usually not constant.)

1. Suppose that the price of a large jar of peanut butter in 1995 was $3.49.

   a. Using an annual inflation rate of 5 percent, find what the price of this jar of peanut butter will be in the year 2010.

   b. Use your work from part a to come up with a formula that will find the price of the peanut butter *N* years after 1995.

In many situations, the value of an item decreases over time. For example, a car that was purchased new for $12,000 five years ago may be worth only $5,000 today. This is called **depreciation.**

If an item depreciates at the rate of 10 percent per year, then the item loses 10 percent of its value each year. That is, at the end of each year, its value is 10 percent less than it was at the start of that year.

2. A fitness club purchases a treadmill for $4800. Because of heavy use, the treadmill depreciates 15% per year.

   a. How much will the treadmill be worth in 10 years?

   b. Use your work in part a to come up with a rule for finding the value of the treadmill after *T* years.

   c. When will the treadmill be worth nothing?

3. What do these problems have to do with *All About Alice?* Describe "Alice" situations that would fit the rules that you found in Questions 1 and 2.

# *More About Rallods*

## *Part I: Counting Rallods*

In *Homework 6: Rallods in Rednow Land,* you studied a situation that involved the powers of 2. To solve that problem, the wise advisor in Rednow Land might have liked to have an easy way to find the sum of such powers. Perhaps you can help.

1. Find a general formula, in terms of $n$, for the sum $1 + 2 + 4 + \cdots + 2^n$. You may want to start by investigating some specific examples, choosing small values for $n$.

## *Part II: Geometric Sequences*

The sequence of powers of 2—that is, 1, 2, 4, 8, 16, and so on—is an example of a **geometric sequence.** A geometric sequence is any sequence of numbers in which each term is a fixed multiple of the previous term. In this example, the multiplier is 2, because each term is twice the one before it.

*Continued on next page*

The multiplier can be any number. For example, if the multiplier is 3 and the first term of the sequence is 1, then the sequence is $1, 3, 9, 27, 81$, and so on.

A geometric sequence can have any number as its first term. For example, if the first term is 12 and the multiplier is $\frac{1}{2}$, then the sequence is $12, 6, 3, 1\frac{1}{2}$, and so on.

2. To get started working with these sequences, develop a general formula for finding a given term of a geometric sequence if you know the first term and the multiplier. For instance, if the first term is $a$ and the multiplier is $r$, what is the fourth term? The tenth term? The hundredth term? The $n$th term?

3. Consider sums of terms of geometric sequences that begin with 1.

   a. Examine sums of the form $1 + 3 + 3^2 + \cdots + 3^k$ for different values of $k$. Find a formula for such a sum in terms of $k$.

   b. Look at examples using different multipliers, and try to find formulas similar to those in Question 1 and Question 3a.

   c. Find a general formula for a sum of the form $1 + r + r^2 + \cdots + r^n$. Your answer should be an expression in terms of $r$ and $n$.

4. Consider sums for general geometric sequences, using $a$ to represent the first term and $r$ to represent the multiplier. Your goal is to find a general formula, in terms of $a$, $r$, and $n$, for the sum $a + ar + ar^2 + \cdots + ar^n$. (*Hint:* Think about how such a sum compares with the corresponding sum for the sequence with the same multiplier but whose first term is 1.)

# *Exponential Graphing*

1. Consider the functions $f$, $g$, and $h$ defined by these three equations.

   $$f(x) = 2^{(x^2)}$$

   $$g(x) = 2^{2x}$$

   $$h(x) = \left(2^x\right)^2$$

   Investigate the graphs of these three functions, and discuss the ways in which each graph is the same as or different from the other two.

2. Use an In-Out table to graph the function given by the equation $y = -(2^{-0.5x})$.

3. Explain why the function whose equation is $y = 2^{-x}$ has the same graph as the function whose equation is $y = \left(\frac{1}{2}\right)^x$.

4. Graph the function whose equation is $y = (-3)^x$ using only integer values for $x$. Describe and explain the behavior of this graph.

# Basic Exponential Questions

The numerical value of an expression like $a^b$ depends on both the base and the exponent. In these problems, you will explore some patterns involved in varying either the base or the exponent, or both.

Be sure to explain your conclusions using examples.

1. If $n < 0$, which is larger, $2^n$ or $3^n$?

2. If $X < 0$, which is larger, $1^{3X}$ or $1^{3 + X}$?

3. Consider the two expressions $2^{5m}$ and $2^{m - 1}$.

   a. For which values of $m$ is $2^{5m} > 2^{m - 1}$?

   b. For which values of $m$ is $2^{5m} < 2^{m - 1}$?

   c. For which values of $m$ is $2^{5m}$ equal to $2^{m - 1}$?

4. If $X$ and $Y$ represent the same number, then the expressions $X^Y$ and $Y^X$ will certainly give the same result. But what if $X$ and $Y$ are not equal? Find out what you can about solutions to the equation $X^Y = Y^X$ in which $X$ and $Y$ are not equal.

# Alice's Weights and Measures

When Alice was growing so that her head reached into space, she learned some lessons about approximation.

She decided to visit a star that was 1 quadrillion miles away. (*One quadrillion* means $10^{15}$.) So she adjusted her height to 1 mile, and then ate 15 ounces of base 10 cake. (She didn't have any problem eating that much, because she was a mile tall to begin with.)

Much to her surprise, she was way off her goal when she finished eating. She looked down and saw some cake crumbs lying at her feet. (Her vision was excellent.)

It turned out that she had dropped some of her cake. She had her friend the Mad Hatter weigh the crumbs and discovered that she had actually eaten only 14.99 ounces (instead of exactly 15 ounces).

1. Without using a calculator, guess how far Alice was from the star.

2. Now calculate how far Alice really was from the star.

3. What percentage of Alice's 15 ounces of cake was dropped as crumbs?

4. By what percentage of 1 quadrillion miles did Alice miss her goal?

*Continued on next page*

5. Suppose Alice had not spilled any cake crumbs, but had been a little careless when she measured her initial height. Specifically, suppose that she was really 0.99 miles tall, rather than 1 mile, but that she did eat exactly 15 ounces of base 10 cake.

   a. Without using a calculator, guess how far Alice was from the star in this case.

   b. Now calculate how far Alice really was from the star.

   c. By what percentage was Alice's initial height measurement off from her estimate of 1 mile?

   d. By what percentage of 1 quadrillion miles did Alice miss her goal?

*A Little Shakes a Lot*

One of the most familiar uses of logarithms (at least in some places) is the **Richter scale,** which is a numerical way of describing the size of an earthquake. Somewhat simplified, the equation used to get the Richter scale number of an earthquake is

$$R = \log_{10} a$$

where $R$ is the Richter scale number and $a$ is the amplitude or amount of the ground motion, as measured on a seismograph.

In order to give intuitive meaning to Richter scale numbers, you need to have some points of reference. For example, an earthquake that measures 4.0 on the Richter scale is barely perceptible outside its immediate center. In contrast, the great San Francisco earthquake of 1906 measured 8.3 on the Richter scale.

1. The Richter scale numbers make it sound as if the 1906 earthquake was only about twice as big as an earthquake that can hardly be felt. But, in fact, that isn't the case at all. Show this by figuring out the answers to these questions.

   a. How many times as much ground motion does an 8.3 quake have compared to a 4.0 quake?

   b. What Richter measurement would represent a quake that has twice the ground motion of one that measures 4.0 on the Richter scale?

   c. What Richter measurement would represent a quake that has half the ground motion of one that measures 8.3 on the Richter scale?

2. The 1989 Loma Prieta earthquake in California was measured at about 7.1 on the Richter scale. In numerical terms, how did the amount of its ground motion compare to that of the 1906 earthquake?

# Who's Buried in Grant's Tomb?

There's a silly old riddle that asks, "Who's buried in Grant's tomb?" Based on this riddle, people sometimes use the phrase "Grant's tomb question" to describe any problem that contains its own answer.

(Another example is the question, "What was the color of George Washington's white horse?")

1. Here are some questions about exponents and logarithms that might be called Grant's tomb questions. Be sure to explain your answers.

   a. What is the cube root of $17^3$?

   b. What is the value of $\log_5(5^8)$?

   c. What is the value of $7^{(\log_7 83)}$?

   d. For what value of $x$ is $\log_x(2^{11})$ equal to 11?

   e. How can you simplify the expression $\left(\sqrt[6]{162}\right)^6$?

2. Make up some Grant's tomb questions of your own. (They don't have to involve exponents and logarithms.)

By the way, the original Grant's tomb riddle is actually a trick question. The elaborate monument in New York City contains both the body of Ulysses S. Grant—the eighteenth president of the United States—and that of his wife, Julia Grant. Most people don't realize that she's also buried there.

# *Very Big and Very Small*

For this activity, you need to think of two situations in which there are numerical questions you would like to investigate.

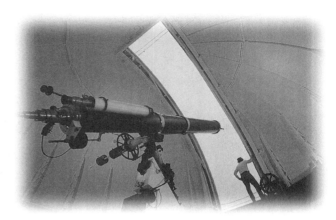

- The first situation should involve *very large* numbers.

- The second situation should involve *very small* numbers.

You can use the examples in *Big Numbers* for ideas.

Write a report on each of your situations. You can use reference books or actual measurements and estimates to get the data for your reports.

*Note:* You may want to check with your teacher about the suitability of your situations and numerical questions.

# Glossary

This is the glossary for all five units of IMP Year 2.

*Absolute growth*  The growth of a quantity, usually over time, found by subtracting the initial value from the final value. Used in distinction from **percentage growth.**

*Additive law of exponents*  The mathematical principle which states that the equation

$$A^B \cdot A^C = A^{B+C}$$

holds true for all numbers *A, B,* and *C* (as long as the expressions are defined).

*Altitude of a parallelogram or trapezoid*  A line segment connecting two parallel sides of the figure and perpendicular to these two sides. Also, the length of such a line segment. Each of the two parallel sides is called a **base** of the figure.

Examples: Segment $\overline{KL}$ is an altitude of parallelogram *GHIJ,* with bases $\overline{GJ}$ and $\overline{HI}$ and segment $\overline{VW}$ is an altitude of trapezoid *RSTU,* with bases $\overline{RU}$ and $\overline{ST.}$

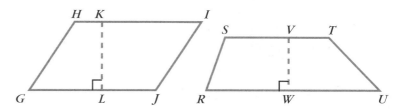

| | |
|---|---|
| *Altitude of a triangle* | A line segment from any of the three vertices of a triangle, perpendicular to the opposite side or to an extension of that side. Also, the length of such a line segment. The side to which the perpendicular segment is drawn is called the **base** of the triangle and is often placed horizontally. |

Example: Segment $\overline{AD}$ is an altitude of triangle *ABC*. Side *BC* is the base corresponding to this altitude.

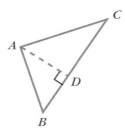

| | |
|---|---|
| *Base* | The side of a triangle, a parallelogram, or a trapezoid to which an altitude is drawn. For **base of a prism**, see *The World of Prisms* in the unit *Do Bees Build It Best?* |
| *Chi-square statistic* | A number used for evaluating the statistical significance of the difference between observed data and the data that would be expected under a specific hypothesis. The chi-square ($\chi^2$) statistic is defined as a sum of terms of the form |

$$\frac{(\text{observed} - \text{expected})^2}{\text{expected}}$$

with one term for each observed value.

| | |
|---|---|
| *Composite number* | A counting number having more than two whole-number divisors. |
| *Cosecant* | See *A Trigonometric Summary* in *Do Bees Build It Best?* |
| *Cosine* | See *A Trigonometric Summary* in *Do Bees Build It Best?* |
| *Cotangent* | See *A Trigonometric Summary* in *Do Bees Build It Best?* |
| *Dependent equations* | See **system of equations.** |

*Distributive property*       The mathematical principle which states that the equation $a(b + c) = ab + ac$ holds true for all numbers $a$, $b$, and $c$.

*Edge*       See **polyhedron.**

*Equivalent equations (or inequalities)*       A pair of equations (or inequalities) that have the same set of solutions.

*Equivalent expressions*       Algebraic expressions that give the same numerical value no matter what values are substituted for the variables.

Example: $3n + 6$ and $3(n + 2)$ are equivalent expressions.

*Expected number*       The value that would be expected for a particular data item if the situation perfectly fit the probabilities associated with a given hypothesis.

*Face*       See **polyhedron.**

*Factoring*       The process of writing a number or an algebraic expression as a product.

Example: The expression $4x^2 + 12x$ can be factored as the product $4x(x + 3)$.

*Feasible region*       The region consisting of all points whose coordinates satisfy a given set of constraints. A point in this set is called a **feasible point.**

*Geometric sequence*       A sequence of numbers in which each term is a fixed multiple of the previous term.

Example: The sequence $2, 6, 18, 54, \ldots$, in which each term is 3 times the previous term, is a geometric sequence.

*Hypothesis*       Informally, a theory about a situation or about how a certain set of data is behaving. Also, a set of assumptions used to analyze or understand a situation.

*Hypothesis testing*       The process of evaluating whether a hypothesis holds true for a given population. Hypothesis testing usually involves statistical analysis of data collected from a sample.

| | |
|---|---|
| *Inconsistent equations* | See **system of equations.** |
| *Independent equations* | See **system of equations.** |
| *Inverse trigonometric function* | Any of six functions used to determine an angle if the value of a trigonometric function is known. |

Example: For $x$ between 0 and 1, the inverse sine of $x$ (written $\sin^{-1}x$) is defined to be the angle between $0°$ and $90°$ whose sine is $x$.

| | |
|---|---|
| *Lateral edge or face* | See *The World of Prisms* in *Do Bees Build It Best?* |
| *Lateral surface area* | See *The World of Prisms* in *Do Bees Build It Best?* |
| *Law of repeated exponentiation* | The mathematical principle which states that the equation |

$$\left(A^B\right)^C = A^{BC}$$

holds true for all numbers $A$, $B$, and $C$ (as long as the expressions are defined).

*Linear equation*   For two variables, an equation whose graph is a straight line. More generally, an equation stating that two linear expressions are equal.

*Linear expression*   For a single variable $x$, an expression of the form $ax + b$, where $a$ and $b$ are any two numbers, or any expression equivalent to an expression of this form. For more than one variable, any sum of linear expressions in those variables (or an expression equivalent to such a sum).

Example: $4x - 5$ is a linear expression in one variable; $3a - 2b + 7$ is a linear expression in two variables.

*Linear function*   For functions of one variable, a function whose graph is a straight line. More generally, a function defined by a linear expression.

Example: The function $g$ defined by the equation $g(t) = 5t + 3$ is a linear function in one variable.

*Linear inequality*

An inequality in which both sides of the relation are linear expressions.

Example: The inequality $2x + 3y < 5y - x + 2$ is a linear inequality.

*Linear programming*

A problem-solving method that involves maximizing or minimizing a linear expression, subject to a set of constraints that are linear equations or inequalities.

*Logarithm*

The power to which a given base must be raised to obtain a given numerical value.

Example: The expression $\log_2 28$ represents the solution to the equation $2^x = 28$. Here, "log" is short for *logarithm,* and the whole expression is read "log, base 2, of 28."

*Net*

A two-dimensional figure that can be folded to create a three-dimensional figure.

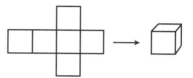

Example: The figure on the left is a net for the cube.

*Normal distribution*

See *Normal Distribution and Standard Deviation Facts* in *Is There Really a Difference?*

*Null hypothesis*

A "neutral" assumption of the type that researchers often adopt before collecting data for a given situation. The null hypothesis often states that there are no differences between two populations with regard to a given characteristic.

*Order of magnitude*

An estimate of the size of a number based on the value of the exponent of 10 when the number is expressed in scientific notation.

Example: The number 583 is of the second order of magnitude because it is written in scientific notation as $5.83 \cdot 10^2$, using 2 as the exponent for the base 10.

*Parallelogram*     A quadrilateral in which both pairs of opposite sides are parallel.

Example: Polygons *ABCD* and *EFGH* are parallelograms.

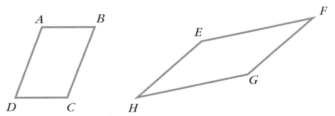

*Percentage growth*     The proportional rate of increase of a quantity, usually over time, found by dividing the absolute growth in the quantity by the initial value of the quantity. Used in distinction from **absolute growth.**

*Polygon*     A closed two-dimensional figure consisting of three or more line segments. The line segments that form a polygon are called its sides. The endpoints of these segments are called **vertices** (singular: **vertex**).

Examples: All the figures below are polygons.

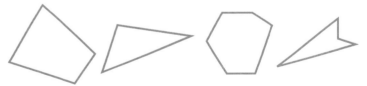

*Polyhedron*     A three-dimensional figure bounded by intersecting planes. The polygonal regions formed by the intersecting planes are called the **faces** of the polyhedron, and the sides of these polygons are called the **edges** of the polyhedron. The points that are the vertices of the polygons are also **vertices** of the polyhedron.

Example: The figure below shows a polyhedron. Polygon *ABFG* is one of its faces, segment $\overline{CD}$ is one of its edges, and point *E* is one of its vertices.

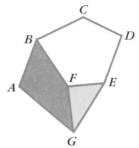

| | |
|---|---|
| *Population* | A set (not necessarily of people) involved in a statistical study and from which a sample is drawn. |
| *Prime factorization* | The expression of a whole number as a product of prime factors. If exponents are used to indicate how often each prime is used, the result is called the **prime power factorization.** |
| | Example: The prime factorization for 18 is $2 \cdot 3 \cdot 3$. The prime power factorization for 18 is $2^1 \cdot 3^2$. |
| *Prime number* | A counting number that has exactly two whole-number divisors, 1 and itself. |
| *Prism* | A type of polyhedron in which two of the faces are parallel and congruent. For details and related terminology, see *The World of Prisms* in *Do Bees Build It Best?* |
| *Profit line* | In the graph used for a linear programming problem, a line representing the number pairs that give a particular profit. |
| *Pythagorean theorem* | The principle for right triangles which states that the sum of the squares of the lengths of the two legs equals the square of length of the hypotenuse. |
| | Example: In right triangle *ABC* with legs of lengths *a* and *b* and hypotenuse of length *c,* the Pythagorean theorem states that $a^2 + b^2 = c^2$. |

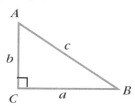

| | |
|---|---|
| *Right rectangular prism* | See *The World of Prisms* in *Do Bees Build It Best?* |
| *Sample* | A selection taken from a population, often used to make conjectures about the entire population. |

| | |
|---|---|
| *Sampling fluctuation* | Variations in data for different samples from a given population that occur as a natural part of the sampling process. |
| *Scientific notation* | A method of writing a number as the product of a number between 1 and 10 and a power of 10.<br><br>Example: The number 3158 is written in scientific notation as $3.158 \cdot 10^3$. |
| *Secant* | See *A Trigonometric Summary* in *Do Bees Build It Best?* |
| *Sine* | See *A Trigonometric Summary* in *Do Bees Build It Best?* |
| *Standard deviation* | See *Normal Distribution and Standard Deviation Facts* in *Is There Really a Difference?* |
| *Surface area* | The amount of area that the surfaces of a three-dimensional figure contain. |
| *System of equations* | A set of two or more equations being considered together. If the equations have no common solution, the system is **inconsistent.** Also, if one of the equations can be removed from the system without changing the set of common solutions, that equation is **dependent** on the others, and the system as a whole is also **dependent.** If no equation is dependent on the rest, the system is **independent.**<br><br>In the case of a system of two linear equations with two variables, the system is *inconsistent* if the graphs of the two equations are distinct parallel lines, *dependent* if the graphs are the same line, and *independent* if the graphs are lines that intersect in a single point. |
| *Tangent* | See *A Trigonometric Summary* in *Do Bees Build It Best?* |
| *Tessellation* | Often, a pattern of identical shapes that fit together without overlapping. |

| | |
|---|---|
| *Trapezoid* | A quadrilateral in which one pair of opposite sides is parallel and the other pair is not.<br><br>Example: Polygons *KLMN* and *PQRS* are trapezoids. |

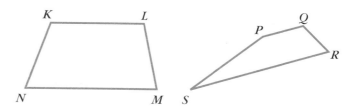

| | |
|---|---|
| *Trigonometry* | For a right triangle, the study of the relationships between the acute angles of the triangle and the lengths of the sides of the triangle. For details, see *A Trigonometric Summary* in the unit *Do Bees Build It Best?* |
| *Vertex* | See **polygon** and **polyhedron.** |
| *x-intercept* | A point where a graph crosses the *x*-axis. Sometimes, the *x*-coordinate of that point. |
| *y-intercept* | A point where a graph crosses the *y*-axis. Sometimes, the *y*-coordinate of that point. |

# Photographic Credits

## Interior Photography

**3** Lincoln High School, Lori Green; **19** Santa Cruz High School, Kevin Drinkard, Lynne Alper; **38** Foothill High School, Cheryl Dozier; **67** Brookline High School, Priscilla Burbank-Schmitt, Carla Oblas; **72** Lumina Designworks, Terry Lockman; **77** Hillary Turner; **79** Fresno High School, Dave Calhoun; **107** Santa Cruz High School, Kevin Drinkard, Lynne Alper; **116** Santa Maria High School, Mike Bryant; **130** Santa Cruz High School, Kevin Drinkard, Lynne Alper; **136** Comstock © 1993; **145** Comstock © 1995; **156** Santa Cruz High School, Kevin Drinkard, Lynne Alper; **168** Hillary Turner; **173** Foothill High School, Cheryl Dozier; **197** Santa Maria High School, Mike Bryant; **200** The Image Bank; **201** Santa Maria High School, Mike Bryant; **225** Watsonville High School, Grace Patiño; **241** Pleasant Valley High School, Mike Christensen; **249** Foothill High School, Cheryl Dozier; **260** Comstock © 1991; **270** Live Oak High School, LeighAnn McCready, Lynne Alper; **271** Hillary Turner; **304** Hillary Turner; **311** Colton High School, Sharon Taylor; **325** West High School, Dean Medek; **335** Comstock © 1995; **336** Comstock © 1996; **339** Pleasant Valley High School, Mike Christensen; **347** Capuchino High School, Chicha Lynch; **354** Pleasant Valley High School, Mike Christensen; **370** Comstock © 1996; **379** Napa High School, Steve Hansen, Lynne Alper; **382** Archive Photos; **386** Comstock © 1993; **387** San Lorenzo Valley High School, Dennis Cavaillé, Lynne Alper; **400** San Lorenzo Valley High School, Dennis Cavaillé; **412** Truman High School, Regina Schwartz, Arlene DeSimone, Lynne Alper; **419** The Image Bank; **432** The Image Bank; **433** The Image Bank; **434** Tony Stone Images; **434** The Image Bank

## Cover Photography and Cover Illustration

**Background** © Tony Stone Worldwide **Top left to bottom right** From *Alice in Wonderland* by Lewis Carroll; Hillary Turner; Hillary Turner; © Image Bank

## Front Cover Students

Colin Bjorklund, Liana Steinmetz, Sita Davis, Thea Singleton, Jenée Desmond, Jennifer Lynn Anker, Lidia Murillo, Keenzia Budd, Noel Sanchez, Seogwon Lee, Kolin Bonet (photographed by Hillary Turner)